Lectures on
GAS CHROMATOGRAPHY
1966

Lectures on
GAS CHROMATOGRAPHY
1966

Based in Part on Lectures Presented at the Eighth Annual
Gas Chromatography Institute, Held at Canisius College

Edited by

Leonard R. Mattick

Food Research Laboratory
New York State Agricultural Experiment Station
Cornell University, Geneva, New York

and

Herman A. Szymanski

Department of Chemistry
Canisius College, Buffalo, New York

ℚ PLENUM PRESS · NEW YORK · 1967

Library of Congress Catalog Card Number 61-15520

ISBN 978-1-4684-9104-3   ISBN 978-1-4684-9102-9 (ebook)
DOI 10.1007/978-1-4684-9102-9

© 1967 Plenum Press
Softcover reprint of the hardcover 1st edition 1967

A Division of Plenum Publishing Corporation
227 West 17 Street, New York, N. Y. 10011
All rights reserved

FOREWORD

Many of the papers in this volume are based on lectures presented at the Eighth Annual Gas Chromatography Institute held at Canisius College, Buffalo, New York, April 11-15, 1966, and at the Pesticide Residue Analysis Workshop conducted at the Institute by Leonard R. Mattick. These lectures are here supplemented by several papers prepared, at the invitation of the editors, by authors whose work, though not presented at the Institute, helps to illustrate current progress in gas chromatography.

CONTENTS

Characteristics of the Phosphate Sensitive Sodium Sulfate Modified Hydrogen Flame Ionization Detector

Royce E. Johnson

Barber-Colman Company
Rockford, Illinois

Highly selective and sensitive gas chromatograph ionization detectors have become indispensable in this age of pesticides and herbicides. The electron attachment detector, for example, has aided chemical manufacturers in developing many effective, safe-to-use pesticides. Regulatory agencies have found the electron attachment detector indispensable for their pesticide residue examinations.

After several years of searching for a selective and equally sensitive detector for phosphated pesticides, one has been reported upon by Giuffrida [1] of the United States Food and Drug Administration Department of Chemistry. Her report in the "Journal of Official Agricultural Chemists" describes this detector and lucidly presents the results of valuable original exploratory work with it.

Giuffrida found that various hydrogen flame detectors, when modified, gave comparable results for organophosphate pesticides. Her report contains information on the essential characteristics of the detector and the sodium salt modification thereof. It also contains information on the relation of hydrogen flow to response. Her table of relative retention

1

times on a DC-200 column for 21 phosphated pesticides is also a contribution to work in this field.

The present discussion is more concerned with operational characteristics of the detector and methodology than it is with the columns. Characteristics of the sodium salt electrode modification and the conventional flame detectors are compared to some extent. This will assist a flame detector operator when he first uses the sodium salt coated electrode and, it is hoped, also will be helpful to those using the sodium thermionic emission detector without prior experience with the flame detector. Flame detector theory is reviewed and several chromatograms are shown in this comparison.

APPARATUS

For this report a helically coiled platinum wire electrode, as used by Giuffrida, was mounted on the burner tip of a Barber-Colman Company Model 5121 Hydrogen Flame Ionization Detector. The detector was placed in the Model 5360 Pesticide Analyzer instead of the electron attachment detector normally supplied with this instrument. A 6 ft by 4 mm i.d. coiled glass column with 5% SF-96 on 90-100 mesh ABS support was used.

A view of the detector base with the sodium salt coated platinum electrode on the burner tube is shown in Fig. 1. The chimney with its side arms for ignition loop and anode support and electrical connections has been removed for the photograph, but removal is unnecessary for installation of the electrode. Effluent and hydrogen burn as the mixture flows from the burner tube up through the helix.

GENERAL THEORY

Ionization takes place in the combustion zone. In the hydrogen flame detector, ionization efficiency is between 0.0005 and 0.001%. Only the few molecules of the sample that acquire

Fig. 1. Hydrogen flame ionization detector base with sodium sulfate coated electrode on flame tube tip.

sufficient energy in the flame become ionized [2,3]. As a first approximation for compounds of carbon and hydrogen, the magnitude of ionized current is proportional to the number of carbon atoms involved.

Organic materials containing oxygen or halogens are ionized in a hydrogen flame even less efficiently. Sternberg [2] relates the decrease in response quantitatively to the number and position of halogen and oxygen atoms in the molecules.

For the hydrogen flame ionization detector, Sternberg reported contributions in units of ionization for various functional groups and molecular bondings. In olefinic, acetylenic, carbonyl, and nitrile molecular structures, the carbon number contributions are 0.95, 1.2, 0.0, and 0.3, respectively. Oxygen contributions range from -1.0 in ether through -0.6 for alcohols to -0.25 for tertiary alcohols. Chlorine contributions range from -0.12 to +0.05 in aliphatic and olefinic compounds.

The sodium salt coated electrode in the hydrogen flame ionization detector superimposes an additional source of ions on the basic or background hydrogen flame ionization. The base or standing current when the sodium electrode is used is so much higher than the base current for the flame that the latter may be practically negligible in comparison. This relation exists with normal column bleed levels. In the modified flame ionization detector, the ionizing efficiency of the thermionic sodium atmosphere for typical phosphated pesticides is between two and three orders of magnitude higher than that of the hydrogen flame ionization detector. Giuffrida [1] reports that for a compound containing 6 chlorine atoms, the response was enhanced about 2000% by the sodium salt coating.

For hydrocarbons the ionizing efficiency of the two detectors is approximately the same for equal hydrogen combustion rates; however, the sodium electrode detector requires somewhat higher hydrogen flow rate.

PERFORMANCE

In the typical hydrogen flame ionization detector, hydrogen flow in excess of the optimum gives less response to organic materials, as is shown in Fig. 2, where the materials consist of stationary phase vapors. The extent to which the standing current decreases as the hydrogen flow exceeds the optimum varies with conditions. If there are organic vapors in the hydrogen, whether coming from the cylinder or the plumbing, an increased flow will carry more combustibles per second to the flame. Excessive contamination in the hydrogen will obliterate the valley so that the standing current will increase continuously with the hydrogen flow rate. It is to be noted that at some flow (48 to 53 ml/min for column flows in Fig. 2) sufficient heat is liberated to cause emission of electrons, as for example, the increase in standing current occurring for conditions in Fig. 2 at about 53 ml/min with a column flow of 57 ml/min. Observation of the platinum anode loop reveals increasing incandescence as the hydrogen flow is increased beyond that at the bottom of the valley in the

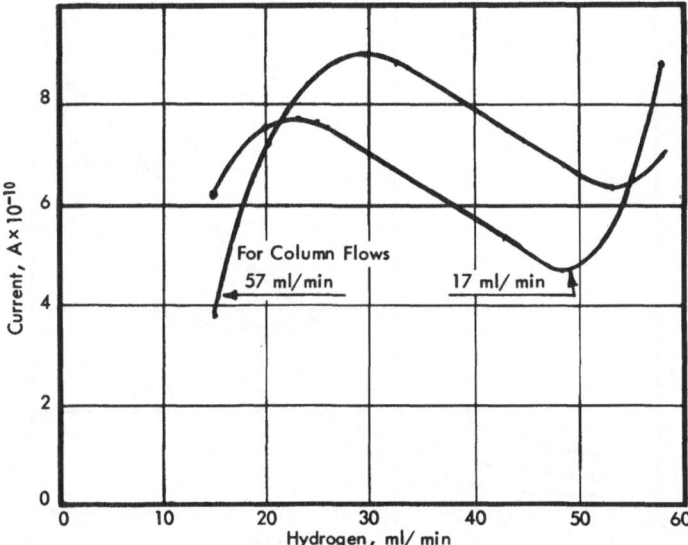

Fig. 2. Hydrogen flame ionization detector background current versus hydrogen flow.

curves of Fig. 2. The ionization current caused by excessive hydrogen flow is not stable, is noisy, and does not usefully contribute to the ionization efficiency of the detector.

Figure 3 shows the relation between hydrogen flow in the sodium thermionic emission detector and the base current, noise, and peak height for a uniform sized sample of parathion. It is to be noted in Fig. 3 that the noise and signal increase more or less similarly for base currents exceeding 10 or 12 times 10^{-10} A. The base and sample peak currents (and therefore the peak area) are definitely influenced by the hydrogen flow rate. At a base current of $5 \cdot 10^{-10}$ A, a 20% change in current is accompanied by approximately an 8% change in peak height as a result of change in hydrogen flow from 32 to 32.3 ml/min. In other words, a 1% change in hydrogen flow causes an 8% change in ionization efficiency. When operating at higher rates of hydrogen flow, the effect of flow variation is even greater.

Detector oven temperature is also much more critical

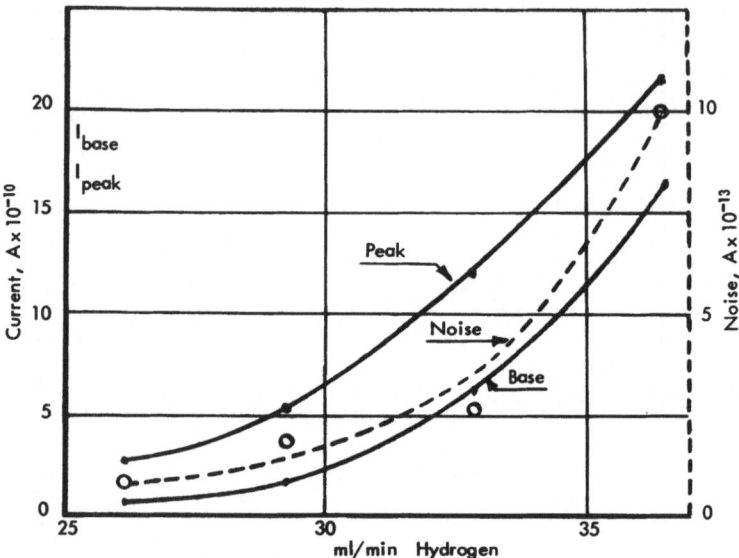

Fig. 3. Sodium modified detector — peak, noise, and background current versus hydrogen flow. Detector, 230°C; Column, 200°; 32 ml/min nitrogen; 20 ng parathion sample.

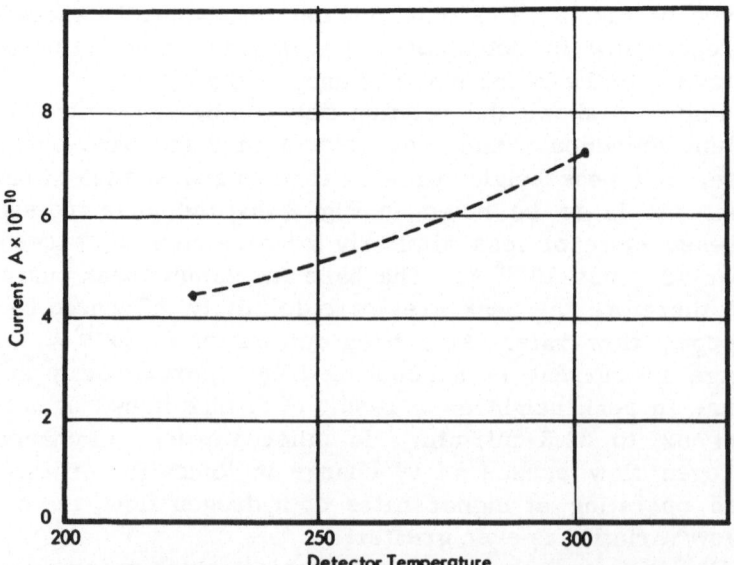

Fig. 4. Sodium modified detector — background current versus detector temperature. Column, 200°C; 17 ml/min nitrogen; constant hydrogen flow.

than for the flame ionization detector. Figure 4 shows that in going from 225 to 310°C, the base current increased some 50%, from 4.4 to $7.2 \cdot 10^{-10}$ A. Temperature coefficient of the hydrogen flame ionization detector is reported by Sternberg [2] to be about 0.1% per degree centigrade.

A chromatogram from a hydrogen flame ionization detector for parathion, malathion, and hexadecane is shown in Fig. 5. Relative responses are tabulated in Table I for this chromatogram, except for the uncertain quantity of malathion caused by its decomposition during this work.

Chromatograms of a commonly used reference pesticide for gas chromatography experimental work, parathion, and another phosphated pesticide known by the acronym EPN are contained in Figs. 6 and 7.

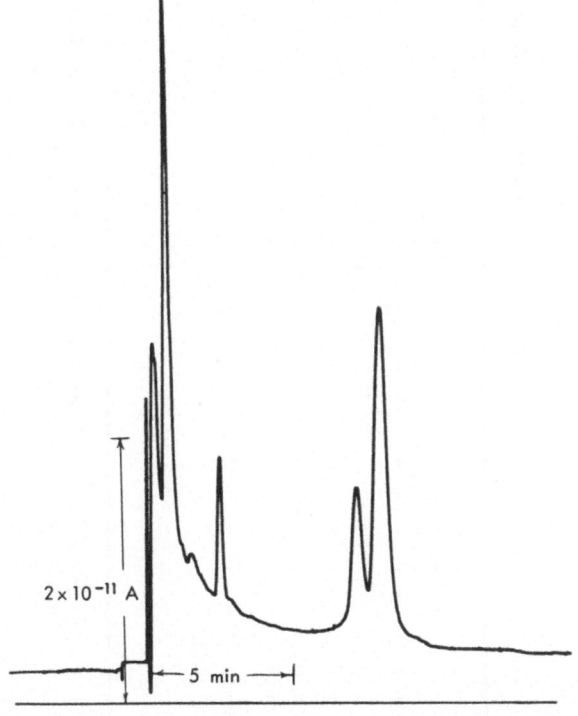

Fig. 5. Hydrogen flame ionization detector chromatogram. Elution sequence: carbon disulfide (0.65 μl) solvent, hexadecane (19 ng), malathion (1950 ng), parathion (65 ng), detector at 230°C.

Table. 1. Relative Response Data

Detector	Relative area per nanogram						Relative peak heights (0.4 µl) (quantity)	
	Hexadecane		Parathion		EPN			
	Nanograms	Relative response	Nanograms	Relative response	Nanograms	Relative response	Acetone	Hexane
Hydrogen flame ionization	19	6	650	1			0.55	1.0*
Sodium salt modification			2	900	2	370	0.55	1.0
			100	740	100	510		
			1000	600	1000	640		

*Ratio of hexane peaks for the two detectors was not determined since this hydrogen flame detector was in another instrument with higher column flow than for the sodium salt modified detector.

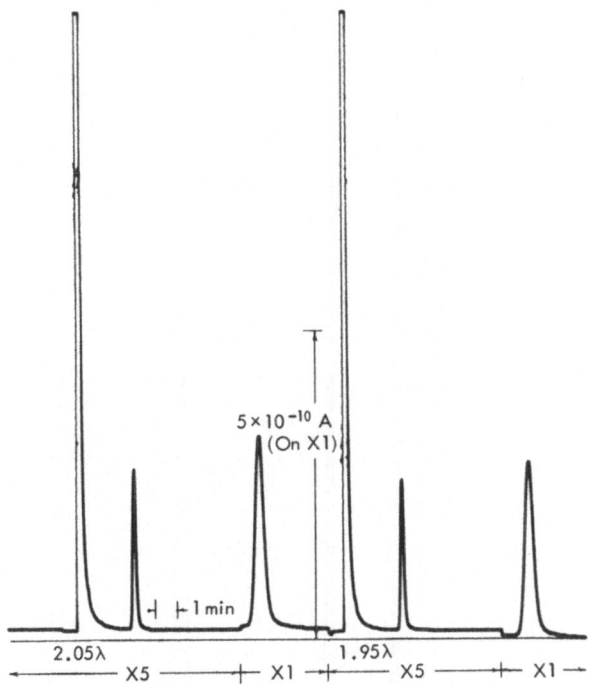

Fig. 6. Sodium modified detector chromatograms. Ten ppm parathion and EPN in
acetone base current, 4 · 10⁻¹⁰A. Detector at 230°C.

It is evident in Fig. 6 that there is good sensitivity for
1-nanogram (ng) samples. With care to control hydrogen flow,
baseline deviations of 20 sec or longer duration due to this
parameter will be no larger than the readily attainable noise
level of about $2.5 \cdot 10^{-13}$ A shown in Fig. 3 for a standing
current of $5 \cdot 10^{-10}$ A.

Extrapolating response for EPN traces, as seems reasonable from Fig. 8, one would estimate on the basis of a minimum usable peak of two times the noise (peak to peak) that
as little as 40 picograms (pg) of EPN could be quantitatively
estimated. Similarly, as little as 4 pg of parathion would be
measurable.

The area per unit of sample varies in a straight line
manner on a log–log plot as the sample size increases. Co-

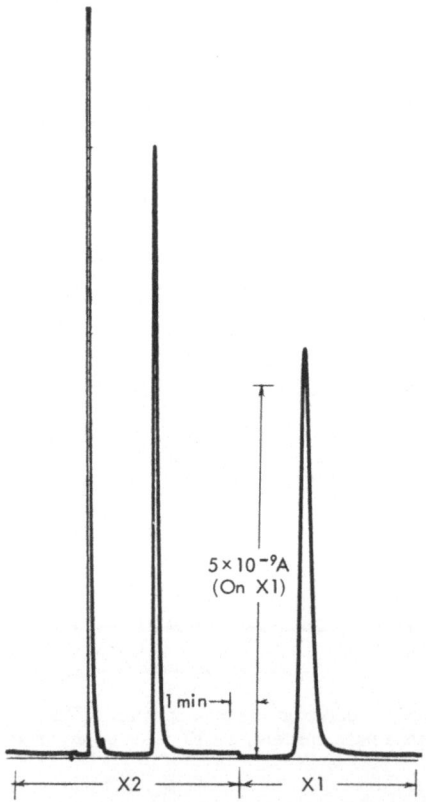

Fig. 7. Sodium modified detector chromatogram. Parathion and EPN (last peak) 208-ng components.

efficients are given in the next section, and data are summarized in Table I.

GENERAL

A factor in the response for larger samples appears to be related to the heat of reaction in the sodium-coated electrode zone. Careful observation of the baseline of Fig. 7 reveals that the standing current is higher after the sample peak than before. This suggests that the electrode temperature has increased during peak elution. Tailing does not appear to be the cause of the higher baseline current; it is too flat for that. During operation, when large samples were chromatographed successively, it was noted that the base current increased, requiring some reduction of the hydrogen flow to restore the base current.

It was observed that solvent peak tailing was related to the base current magnitude. Tailing became more pronounced as the base current was decreased below an optimum level by reducing hydrogen flow. Increasing the hydrogen flow above the optimum caused a negative pen deflection instead of tailing. Under conditions of this work the optimum base was approximately $5 \cdot 10^{-10}$ A. This relation of hydrogen flow to solvent tailing was subsequently observed in other installations.

The relation between base current and response is shown in Fig. 9. Here, as in Fig. 3, it is evident that with currents higher than $5 \cdot 10^{-10}$ to $8 \cdot 10^{-10}$ A no gain in signal-to-noise

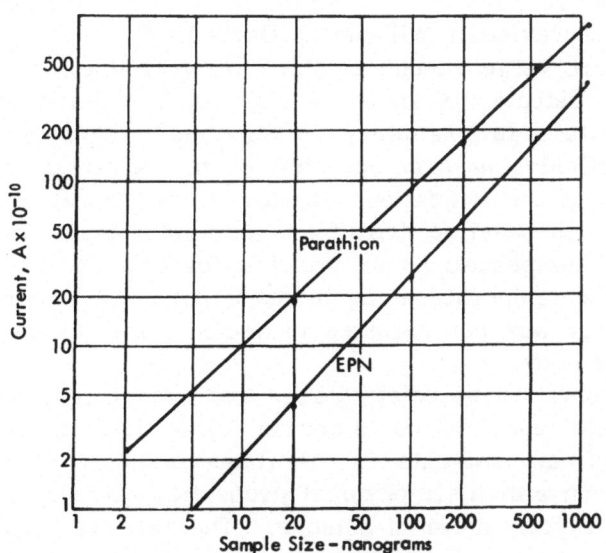

Fig. 8. Peak height versus sample size — Sodium modified detector. Column, 200°C; 32 ml/min nitrogen carrier.

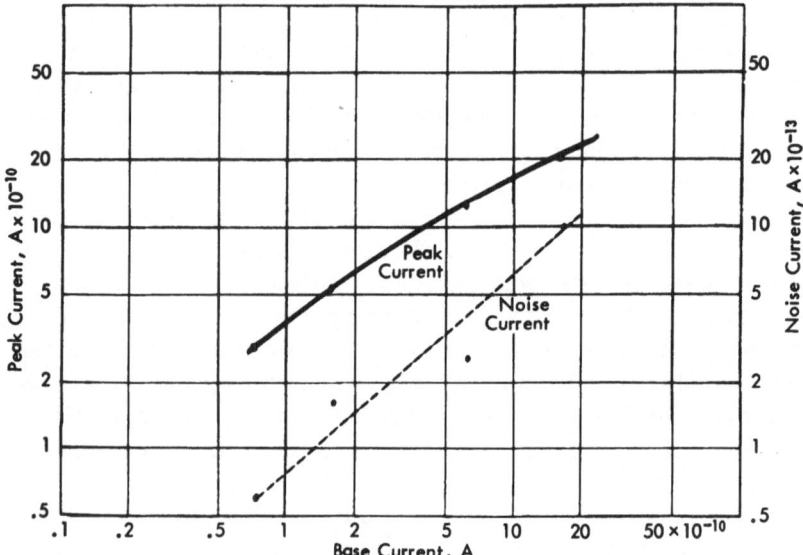

Fig. 9. Peak height and noise versus background current — sample 20 ng parathion; 100 ppm in acetone. Column 200°C; 32 ml/min nitrogen.

ratio can be expected. These figures are very similar to those obtained previously William M. Barbour [4].

Where large enough samples are available, one advantage of this detector is shown in Fig. 10, namely, single column temperature programming is feasible. The response from the pesticides is large enough to permit relatively low amplification of the ionization current. Consequently, it does not show much baseline drift from increasing bleed vapor. The rate of temperature rise was low for this chromatogram to minimize temperature lag in the column. A linear program controller was not required to demonstrate the low apparent baseline shift.

A summary of relative area and peak height data for the materials used is contained in Table I. The peak height ratio of the solvents in the flame detector appears to be consistent with their physical properties, density, and effect of the oxygen atom in acetone. The relative responses for hexane and acetone are the same in the sodium thermionic emission detector as in a flame ionization detector of identical design. This is of significance only in that it shows the response in the sodium detector to be similar to that in a flame detector of the same basic type. The time required to make the comparison under identical hydrogen flame ionization conditions for the normal and for the sodium electrode modification precluded accurate experimental verification of relative response in the two types of detectors.

Since the peak height per nanogram versus nanograms of sample is a straight line (for the range tested) on log–log paper, the data can be expressed by an equation of the form

$$i = bq^m$$

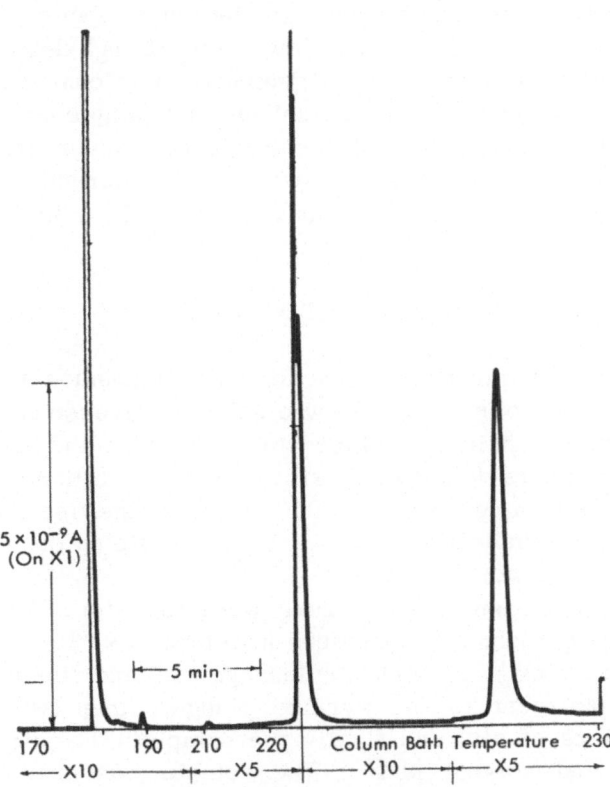

Fig. 10. Sodium modified detector chromatogram for programmed temperature. 5% SF-96, 6 ft × 4 mm ID glass column, 170 to 220°C—sample 1.05 ng parathion, 1.05 ng EPN, acetone solvent.

where i is peak current above base current; m and b are coefficients (m is the slope on log-log paper); and q is quantity
of sample.

For current in units of 10^{-10} amperes and quantity in
nanograms, the coefficients are as follows:

Curve	m	b
Fig. 8, Parathion	0.94	1.18
Barbour (4), Parathion.	1.34	0.046
Fig. 8, EPN.	1.10	0.16

Obviously, the coefficients are dependent upon parameters.
These may range from gas flow rates through detector temperature, polarizing voltage, geometry, and coated electrode
details to factors such as the solvent and sample composition.
Stability and consistency of performance, however, are entirely
adequate for quantitative work when using reference standards
with concentrations comparable to those in the samples.

METHODOLOGY

All of the chromatograms for this discussion were made
on an SF-96 column. SF-96 was selected because of previous
experience with it for halogenated pesticides and because its
oxidized vapors do not leave silica deposit in a flame detector
that is associated with DC-200. Whether the fine silica dust
deposit adversely affects the sodium sulfate on the electrode
was not ascertained.

Sample measurement errors were kept at a ± 3% minimum
by using the technique described in references [5] and [6]. This
technique uses a combination leakage seal and rinsing plug of
solvent adjacent to the syringe plunger, then two or three
microliters of air, and finally the sample. The air isolates
sample and solvent plug. The sample was drawn into the
$10 \, \mu l$ syringe barrel with the lower meniscus no higher than
the $1 \, \mu l$ mark. Its volume could be observed to ± 0.01 μl.
With the plunger held in this position the needle was fully

inserted through the septum of the 260°C inlet, and the plunger was pushed to its limit for a 5-sec count. The plunger was then retracted before the needle was withdrawn so that the solvent vapors would be forced back into the syringe where the remaining quantity of solvent could be measured. Inconsistency of injection technique is indicated by a variable amount of residual solvent if the hold time for the insertion is not too long. Care was used to thoroughly rinse the syringe after each sample and not to wet with the sample any more of the syringe bore than necessary.

Retention of the sample in the syringe barrel during needle penetration of the septum avoids loss of sample by capillary action during penetration. It also minimizes the peak broadening effect of sample vaporization as the needle tip emerges through the septum and moves to the fully inserted position. Diffusion of the sample vapor into the space between septum and carrier gas inlet is avoided.

1. Increase in carrier gas flow decreased the base current even though the SF-96 column was at 240°. Apparently the carrier gas cooled the incandescent electrode. A change from no carrier flow to 40 ml/min through a 210° column reduced the base current from $9 \cdot 10^{-10}$ to $5 \cdot 10^{-10}$ A in a 310°C sodium electrode detector.

2. Somewhat more air for combustion is required than for the hydrogen flame ionization detector. Adequate air is determined by increasing the supply until an appreciable increase in base current (for a given hydrogen flow) is not produced by more air; 10 to 25% more air flow was found helpful.

3. Carbon disulfide solvent produced more response for the solvent than hexane in the sodium salt coated electrode detector under the prevailing conditions for the first few samples. However, it reduced the base current and response to phosphate pesticides. Three successive CS_2 samples totaling 6 μl reduced the base current from $5.5 \cdot 10^{-10}$ to $3 \cdot 10^{-10}$ A. Three hours later the current had recovered to only $4.5 \cdot 10^{-10}$. To restore it to normal at that time more hydrogen was used.

CONCLUSION

Each type of sensitive gas chromatograph detector is responsive to certain parameters. The sodium thermionic emission detector is relatively unresponsive to stationary phase vapors and to normal contaminants in gas supplies. These have no noticeable direct effect on its response to materials for which it is specific. Stable hydrogen flow and detector temperature are essential. Sample size moderately and consistently influences response (area) per unit weight or volume in a moderate and consistent manner.

Response of the sodium thermionic emission detector is from 500 to 1100 times greater for parathion than that of the hydrogen flame detector. Peak response per unit varies with sample size as a power of its size. For the curves analyzed the power ranged between 0.94 and 1.34. As little as 40 pg of EPN and 4 pg of parathion can be measured.

As techniques are developed for preparing suitable extracts and derivatives, this detector will fill a need in biochemical research laboratories working with phosphorous-containing compounds and will have immediate application in pesticide residue work.

REFERENCES

1. L. Giuffrida, "A Flame Ionization Detector Highly Selective and Sensitive to Phosphorous — A Sodium Thermionic Detector," J. Assoc. Off. Agr. Chem. Vol. 47, No. 2, 1964.
2. J.C. Sternberg, W.S. Gallaway, and T.L. Jones, "The Mechanism of Response of Flame Ionization Detectors," in: Gas Chromatography, Third International Symposium, Academic Press, New York, 1961.
3. R.E. Johnson, Lectures on Gas Chromatography, ed. H.A. Szymanski, Plenum Press, New York, 1962, pp. 65–86.
4. W.M. Barbour, (Internal) Report on Sodium Thermionic Detector, Barber-Colman Company, December, 1964.
5. Barber-Colman, Chromatog. Quart., Vol. 1, No. 4, 1961.
6. R.E. Johnson, Practical Suggestions for Obtaining Maximum Performance in Electron Attachment Gas Chromatography, Gas Chromatography Institute, Canisius College, Buffalo, New York, April 21, 1965.

Analysis of Organophosphorus and Organic Iodine Pesticide Residues by Microwave Powered Emission Spectrometry

C. A. Bache and D. J. Lisk

Pesticide Residue Laboratory
Cornell University
Ithaca, New York

Several selective detectors for use in gas chromatographic analysis of organophosphorus insecticides have been developed. The microcoulometric detector [5] and the sodium thermionic detector [6] are well known and are being routinely used for residue analysis. More recently the flame emission [3], the stacked flame detector [7], and the phosphine reduction detector [4] have become commercially available for these compounds.

The true value of a pesticide detector can be adequately assessed only by noting its signal to noise ratio in the presence of extraneous sample substances. A high ratio is needed for rapid, sensitive analysis. A detector that admirably possesses these features was constructed in 1964 [9]. It is based on a combination of gas chromatography and emission spectrometry. The construction of this system has been described in detail [9].

Briefly, the exit from the gas chromatographic column is connected to a very narrow quartz capillary tube. The tube is positioned in the cavity of a microwave generator that

17

couples the tube into the strong microwave field. With argon carrier gas flowing through the column and tube, a spark is introduced into the tube to ignite an intense, bluish plasma, which is then maintained by the continuous microwave energy. Molecules entering the plasma are torn apart by electron (those provided in the spark) bombardment, bonds are broken, and excited monatomic and diatomic fragments are largely produced. These species emit characteristic atomic lines and molecular bands that are focused on the slit of a high-resolution grating spectrometer. With the grating adjusted to monitor the desired line, a selective detector system results. A photomultiplier tube, amplifier, and recorder present the chromatogram with line intensity plotted against retention time. Figure 1 shows the apparatus with Mr. Bache igniting the plasma with a Tesla coil.

Two sensitive atomic lines for phosphorus and iodine were found [9] at 2535.7 and 2062 Å, respectively. These lines have been used for analysis of pesticides containing these elements

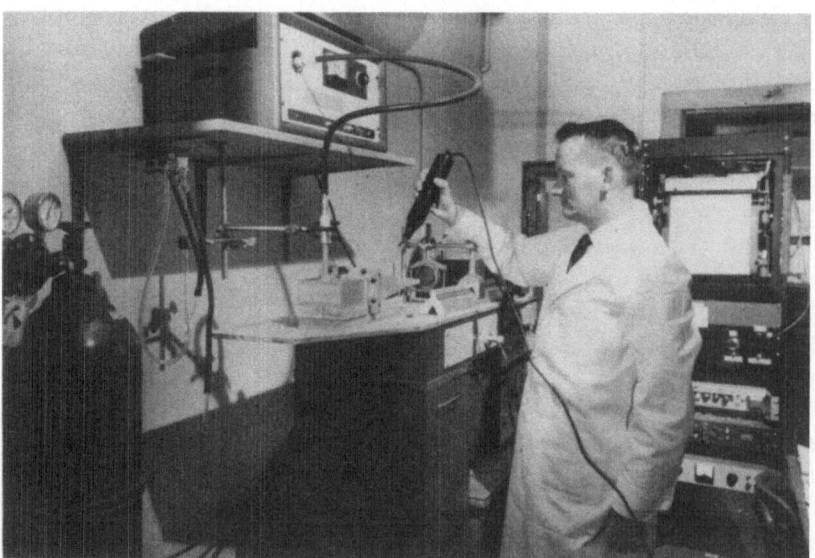

Fig. 1. Photograph of the apparatus with Mr. Bache igniting the plasma with a Tesla coil.

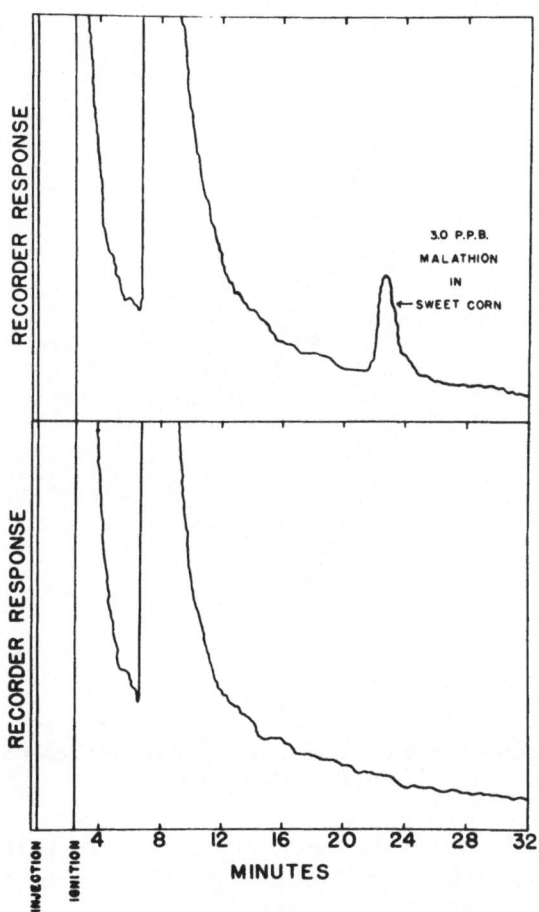

Fig. 2. Chromatograms of 3.0 ppb of malathion added to sweet corn and the control corn.

in a variety of samples [1, 2] with sensitivities of 10^{-11} g of phosphorus or iodine per second. Figure 2 shows chromatograms of 3.0 ppb of malathion [S-(1,2-bis(ethoxycarbonyl) ethyl) O,O-dimethyl phosphorodithioate] in sweet corn and the control corn. Figure 3 shows chromatograms of 0.1 ppm each of ioxynil (3,5-di-iodo-4-hydroxybenzonitrile) herbicide and a possible metabolite, IBA (3,5-di-iodo-4-hydroxybenzoic acid), added to wheat grain and the control. The grain was

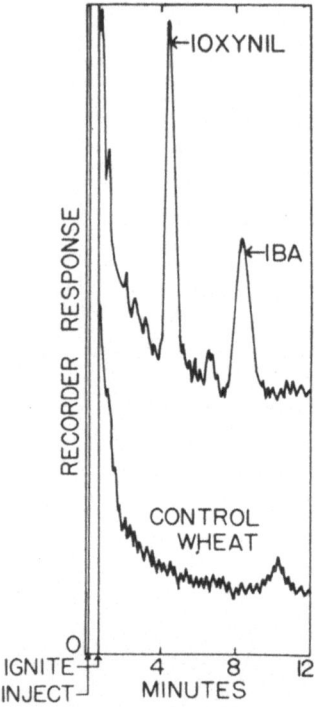

Fig. 3. Chromatograms of 0.1 ppm each of ioxynil and IBA added to wheat grain
and the control wheat.

extracted with benzene, the compounds were partitioned in
sodium bicarbonate, and following acidification they were
extracted into chloroform. This solution was concentrated,
an injection was made representing 1.5 g of grain, and the 2062
Å atomic iodine emission was monitored. Table I lists recoveries of organophosphates and iodinated herbicides in various commodities.

A few words are in order concerning the operation of this
system. The intensity and location of emitted radiation depends greatly upon the nature of the excitation source.
Therefore, one must determine what lines are produced and
which ones would be most desirable for analysis. A spectrum
of the desired compound is therefore first developed by va-
porizing the pure compound into the plasma in a stream of

argon gas and spectrometrically scanning and recording the emitted frequencies. Reference to standard wavelength tables for emitted lines of the elements will aid in ascribing a specific line to a particular elemental emission.

The precise wavelength setting for maximum sensitivity is located by manually setting it approximately (the use of emission standards is helpful). The same amount of the com-

Table I. Recovery of Pesticides Using the Emission Detector

Compound	Sample	Added, ppm	Recovery, %
Organophosphorus insecticides:			
Diazinon	Grapes	0.30	79
Dimethoate	Alfalfa	0.06	106
		0.19	88
	Timothy	0.20	91
	Lettuce	0.03	113
	Milk	0.20	115
	Cow urine	0.20	98
Disyston	Potatoes	0.18	81
	Grapes	0.40	72
Ethion	Soil	0.20	73
Parathion	Bees	0.60	83
	Halibut	0.20	105
	Lettuce	0.20	82
Ronnel	Whole chicken	0.25	71
	Eggs	0.25	90
Organic iodine herbicides:			
Ioxynil	Wheat	0.10	90
	Oats	0.10	78
	Soil	2.00	106
Ioxynil benzoic acid	Wheat	0.10	69
	Oats	0.10	80
	Soil	2.00	97
Mono-iodo ioxynil	Soil	2.00	108

pound is then repeatedly injected while varying the wavelength about 0.2 Å each time in the approximate region. The setting showing the greatest peak height is used.

The particular line chosen for analysis should be governed by the sensitivity required and the presence of nearby interfering lines. Detector selectivity may be judged by injection of increasing amounts of organic compounds of approximately equal molecular weight but not containing the element of interest. High spectrometer resolution will often allow selective analysis when lines that would otherwise interfere are nearby. Carbon to carbon and carbon to nitrogen emission bands commonly appear in emission spectra. There is not much hydrocarbon emission in the deep ultraviolet region.

Further work with this detector system has led to modifications that increase sensitivity to organophosphorus insecticides by a factor of 10 to 15. Its use for determining these insecticides as residues at levels down to 1 ppb with a minimum of preliminary isolation will soon be reported. Selective analysis of other elements in organic compounds has been demonstrated [9].

REFERENCES

1. C.A. Bache and D.J. Lisk, "Determination of Organophosphorus Insecticide Residues Using the Emission Spectrometric Detector," Anal. Chem. 37:1477, 1965.
2. C.A. Bache and D.J. Lisk, "Determination of Iodinated Herbicide Residues and Metabolites by Gas Chromatography Using the Emission Spectrometric Detector," Anal. Chem. 38:783, 1966.
3. S.S. Brody and J.E. Chaney, "Flame Photometric Detector," J. Gas Chromatog. 4:42, 1966.
4. H.P. Burchfield, D.E. Johnson, J.W. Rhoades, and R.J. Wheeler, "Selective Detection of Phosphorus, Sulfur, and Halogen Compounds in the Gas Chromatography of Drugs and Pesticides," J. Gas Chromatog. 3:28, 1965.
5. D.M. Coulson and L.A. Cavanagh, "Automatic Chloride Analyzer," Anal. Chem. 32:1245, 1960.
6. L. Giuffreda, "A Flame Ionization Detector Highly Selective and Sensitive to Phosphorus—A Sodium Thermionic Detector," J. Assoc. Off. Agr. Chem. 47:293, 1964.
7. A. Karmen, "Specific Detection of Halogens and Phosphorus by Flame Ionization," Anal. Chem. 36:1416, 1964.
8. A. Karmen and L. Guiffreda, "Enhancement of the Response of the Hydrogen Flame Ionization Detector to Compounds Containing Halogens and Phosphorus," Nature 201:1204, 1964.

9. A. J. McCormack, S.S.C. Tong, and W.D. Cooke, "Sensitive Selective Gas Chromatography Detector Based on Emission Spectrometry," Anal. Chem. 37:1470, 1965.

Gas Chromatography of Amino Acids

William J. McBride and Jack D. Klingman

Department of Biochemistry
State University of New York at Buffalo
Buffalo, New York

Amino acid gas chromatography procedures were reviewed in 1962 by Burchfield and Storrs [7] and by Littlewood [28]. Since that time additional procedures have been developed, and several workers have used these procedures to analyze the amino acid composition of proteins, blood, and nerve tissue and to resolve optical isomers.

Amino acid gas chromatography procedures are listed in Tables I to IV. The tables are organized to give the reader a brief summary of the procedures and results of the many published works. The original article should be consulted for exact details because only the basic approach to the preparation of derivatives is listed; usually only the one preparation, column packing, and GC condition which gave the best results are listed, although several others may have been investigated.

The analysis of amino acids by gas chromatography has certain advantages over analyses carried out by thin layer or paper chromatographic techniques or by an automatic amino acid analyzer. The gas chromatography equipment is designed for routine, rapid, and automatic separation and quantitation of microgram and submicrogram amounts of materials. Gehrke and Shahrokhi [13] separated the amino acid derivatives obtained from 25 μg of protein hydrolyzate within 60 min. The

Table I. Methyl Ester Derivatives of Amino Acids

Derivative	Preparation, column packing, and gas chromatography conditions	Results
A. Free base 1a. Bayer et al. [1,2]	Amino acids were refluxed for several hours with CH₃OH saturated with HCl. The solution was then treated with 2N NaOH, and the amino acid ester hydrochlorides were extracted with ether. The 3 to 5 m columns were packed with silicone high vacuum grease and sodium caproate on Sterchamol; the oven was operated isothermally between 138 and 191°C with the H₂ carrier flow adjusted to a constant value between 45 and 138 ml/min.	Successfully chromatographed val, norval, leu, norleu, glu, met, and phe; also identified several amino acids in a protein hydrolyzate.
2a. Nicholls et al. [33]	The amino acid methyl ester hydrochlorides were dissolved in CH₃OH and shaken with Dowex-1 anion exchange resin in the hydroxyl form. The column was packed with 2% NPGS on Fluoropak 80; it was operated isothermally between 120 and 180°C with an N₂ flow rate of 60 ml/min.	Only cySH, his, tyr, and try could not be satisfactorily chromatographed in the free base form. The acetate and chloride salts of the amino acid methyl esters were also chromatographed. The acetate salt derivatives gave the same results as those obtained with the free base. The chloride salts of only nine amino acid methyl esters were successfully chromatographed.

B. α-Chloro

1b. Melamed and Renard [32]

The amino acids were treated with a mixture of concentrated HCl and HNO_3 for 1 hr. The resulting α-chloroamino acids were esterified with ethereal diazomethane.

A column consisting of 2 m polyethylene glycol followed by 2 m silicon oil—stearic acid gave the best results. It was operated isothermally at 130°C with H_2 flow rate at 43 ml/min.

An almost 100% yield of α-chloro acids were obtained. Successfully quantitated gly, ala, val, leu, and ileu in a casein hydrolyzate. This technique is limited to simple neutral α-amino acids.

C. α-Hydroxy

1c. Wagner and Rausch [45]

The amino acids were dissolved in 1 N H_2SO_4, and 60% $NaNO_2$ was added at 0°C. The resulting α-hydroxy amino acids were esterified with ethereal diazomethane.

A 183 by 0.6 cm i.d. column containing a silicone impregnated support was used; the column was operated isothermally between 100 and 160°C with a carrier flow of 75 ml/min.

The α-hydroxy methyl esters of arg, phe, his, met, thr, ileu, leu, and val were prepared. The yields were reproducible and ranged from 80% for ileu to 21% for arg and leu. The GC was very poor, and not all of the derivatives were resolved.

D. N-Formyl

1d. Losse et al. [29]

The amino acids were reacted with formic acid in acetic anhydride. The N-formyl amino acids were then esterified with diazomethane in methanol-ether.

The 150 by 0.5 cm i.d. column was packed with 25% high vacuum grease on Kieselguhr; it was operated isothermally at 194°C with a H_2 flow adjusted to a constant value between 2.8 and 4.5 ml/min.

Quantitative yields of 95% or better were reported for gly, ala, val, leu, norval, pro, asp, glu, met, and phe.

Table I. Continued

Derivative	Preparation, column packing, and gas chromatography conditions	Results
E. N-Acetyl 1e. Shlyapnikov and Karpeiskii [39,40]	The amino acids were added to CH₃OH, and HCl was bubbled through until the amino acids dissolved; the reaction was allowed to proceed for 30 min at room temperature. The ester hydrochlorides were then reacted with acetic anhydride for 15 min at room temperature. The column was packed with 1.5% PBDS on Chromosorb W; the temperature was programmed from 80 to 180°C; Argon flow rate was 81 ml/min.	Successfully prepared and chromatographed simple, neutral, amino acid derivatives as well as asp, glu, and met. Many different stationary phases were tried; however, a mixture of the 12 derivatives could not be completely resolved in a single operation. It was also found that the separation of the N-acetyl methyl esters was better than that obtained with N-formyl esters.
F. N-Trifluoroacetyl 1f. Saroff and Karmen [37]	The amino acids were refluxed for 45 min in CH₃OH containing Dowex 50 cation exchange resin in the hydrogen form. The excess methanol was decanted. To the washed resin were added CH₃OH and methyl TFA. This solution was made alkaline with triethylamine and then refluxed for 15 min. The 6 ft by .25 in. column was packed with 22% PEGA on Chromosorb W; it was operated isothermally between 160 and 180°C with an argon flow rate of 80 ml/min.	Suitable single volatile derivatives were obtained for most of the common amino acids. Multiple peaks were obtained for tyr and arg. CyS, his, and try derivatives gave completely negative results.

2f. Wagner and Winkler [44,46]

The amino acids were treated with TFA anhydride at -10 to 0°C for 4 hr. The N-TFA amino acids were then esterified with ethereal diazomethane.

The 150 by 0.6 cm column was packed with a mixed stationary phase of 17.5% Apiezon L and 2.5% sodium caproate on Sterchamol; the column was operated isothermally with a H₂ flow of 38 ml/min.

The N-TFA methyl esters of val, leu, phe, met, and thr were prepared. The GC separation was very poor, and many extraneous peaks were observed.

3f. Ikekawa [22]

The amino acids were refluxed for 3 hr with CH₃OH containing 0.2 meq/ml HCl. The ester hydrochlorides were then dissolved in a methanol solution containing methyl TFA and trimethylamine. The reaction was allowed to proceed for 2 hr at 20°C.

A series column packed with 75 cm 1% NPGS on Gas Chrom P followed by 75 cm 1.5% SE-30 on Gas Chrom P was used; the temperature was programmed at 4°C/min from 95 to 225°C, and the N₂ carrier flow was 60 ml/min.

Suitable derivatives of all common protein amino acids were prepared with the exception of arg, cyS and his. The GC results were excellent. The peaks were very sharp and well resolved. Only the derivatives of ser and met did not completely resolve. A dual column and differential flame were used, and no baseline drift was observed. Better separation and sharper peaks were obtained when the temperature was gradually increased than when operated isothermally.

4f. Cruickshank and Sheehan [8]

The amino acids were refluxed in a dimethyl sulfite-methanolic HCl solution for 30 min. The amino acid methyl esters were then refluxed for 10 min with TFA anhydride.

A second procedure involved refluxing amino acids for 5 min in TFA acid and TFA anhydride. The TFA amino acids were dissolved in CH₃OH and esterified with ethereal diazomethane.

Single volatile derivates were obtained for all the common protein amino acids except arg, which gave two minor components and one major component when methylation was carried out first and no volatile derivative when the acylation step was carried out first. Successfully identified the amino acid residues in a ribonuclease hydrolyzate. Relative

Table I. Continued

Derivative	Preparation, column packing, and gas chromatography conditions	Results
4f. Continued	A 2–3 ft by 0.15 cm i.d. column containing 5% NPGS on Gas Chrom P was used; the temperature was programmed to increase from 65 to 210°C, and the argon flow was 18 ml/min.	peak areas for each amino acid in a standard mixture of 20 were in excellent agreement during 6 repetitions of the procedure in the case where the methylation step was carried out first.
5f. Hagen and Black [18,19]	The amino acids were added to CH_3OH, and thionyl chloride was then added dropwise to the cold solution. The reaction was allowed to proceed for 2 hr at 40°C. The amino acid esters were acylated in TFA anhydride at room temperature for 2 hr. The column lengths varied and were usually packed with 1% Carbowax 20 M or Carbowax 1540 on Diataport S; the temperature was programmed to increase between 2 and 5°C/min, the N_2 flow was 50 or 80 ml/min, and the injection port was at 305°C.	Suitable derivatives of 19 common amino acids were prepared, including arg, try, and his. A mixture of all 19 derivatives could not be completely resolved using any one stationary phase. A linear response was obtained for all 19 derivatives over the range from 1.0 μg to 20 μg using the hydrogen flame ionization detector.
6f. Makisumi and Saroff [30]	The Amino acids were added to CH_3OH which was then saturated with HCl. The solution was refluxed for 1 hr with a slow addition of HCl. The methyl esters were then reacted with TFA anhydride at room temperature for 30 min.	Successfully prepared 21 amino acid derivatives including try, arg, cyS, and his.

6f. Continued

A specially constructed GC apparatus was used. The instrument contained three separate ovens and columns and each was operated under different isothermal conditions. The columns were connected in series; the first was operated between 180 and 210°C, the second at 160°C, and the third at 138°C. The N₂ carrier flow was different in each column. Each was packed with 2% NPGS on Chromosorb W.

7f. Shlyapnikov et al. [42]

The amino acids were added to TFA anhydride at room temperature and allowed to react for 20 min. The TFA amino acids were then esterified with ethereal diazomethane at room temperatures for 30 min.

Glass columns 120 by 0.3 cm packed with 1.5% PBDS on Chromosorb W were used. The temperature was programmed to increase from 80 to 180°C, and the argon flow was 81 ml/min.

Suitable derivatives were prepared for most amino acids except tyr, try, arg, and his. A linear detector response was obtained for the prepared derivatives over a wide weight range using an argon ionization detector. A complete separation of 16 amino acid derivatives was accomplished in a single temperature programmed operation.

G. Dinitrophenyl (DNP)
1g. Pisano et al. [34]

The DNP amino acids were esterified with diazomethane.

The 6 ft by 0.4 cm i.d. glass columns were packed with 1% PhSi or 1% QF-1 on Gas Chrom P and operated isothermally at 194°C with the argon carrier pressure set at a constant value between 12 and 25 psi.

Successfully prepared volatile derivatives for the simple neutral and acidic amino acids. A detector response was not obtained for ser, thr, try, tyr, his, nor for any other basic amino acid.

Table I. Continued

Derivative	Preparation, column packing, and gas chromatography conditions	Results
2g. Landowne and Lipsky [26,27]	The pure DNP amino acids were converted to the methyl esters by reaction with ethereal diazomethane. The column was packed with 3% NPA or NPS on Anakron ABS and operated isothermally at 220°C with a flow of 200 ml/min argon.	Suitable volatile derivatives were obtained for 18 amino acids with the exception of the hydroxy amino acids, his, arg, and try. A linear response was obtained over the range from $4 \cdot 10^{-13}$ to $2 \cdot 10^{-16}$ moles/sec using the electron capture detector that can be adjusted to selectively respond to compounds containing the DNP moiety.

volatile derivatives of a standard mix of protein amino acids were separated by Cruickshank and Sheehan [8] within 70 min, by Gehrke et al. [12] within 60 min, and by Zomzely et al. [57] within 45 min, whereas Benson and Patterson [3], using an automatic amino acid analyzer, required $4\frac{1}{4}$ hr to completely analyze the neutral, acidic, and basic amino acids.

Amino acid analysis by gas chromatography has one distinct disadvantage: the preparation of volatile amino acid derivatives. In order to be of quantitative value, the derivatives must be prepared in high reproducible yields. The procedure should not be very time-consuming, and must be applicable to all protein amino acids to be of full value to investigators in the biological fields. The derivatives must be stable and not so volatile that serious losses occur during preparation.

DERIVATIVES

The wide disparity of structure and chemical properties among the amino acids presents several problems in preparing suitable volatile derivatives.

Volatile derivatives of the simple alphatic amino acids, glycine, alanine, valine, leucine, and isoleucine, are readily prepared by all procedures listed in Tables I to IV. The hydroxy amino acids, serine, threonine, tyrosine, and hydroxyproline, are volatilized more easily after acylation of the free hydroxyl group. Suitable volatile derivatives of arginine, histidine, and tryptophan are the most difficult to prepare. The guanidine group of arginine and the imidazole group of histidine readily form salts under most acylating conditions. The indole nitrogen of tryptophan is difficult to acylate, and unless the second nitrogen is acylated as well, the resulting monoacyl ester will elute from the column only at high temperatures. Lysine is converted to a volatile derivative after acylation of the second amino group; this is readily accomplished under most acylating conditions. Esterification and acylation of cysteine is carried out under a nonoxidizing atmosphere, otherwise it will be converted to cystine, which forms

Table II. Ethyl, Propyl, Isopropyl, and Amyl Ester Derivatives

Derivative	Preparation, column packing, and gas chromatography conditions	Results
I. Ethyl, propyl, and isopropyl esters		
A. N-Acetyl		
1a. Graff et al. [17]	The amino acids were refluxed in a mixture of n-propanol and benzene. The resulting amino acid esters were then acylated with acetic anhydride. The GC column was packed with PEG on Chromosorb W.	The abstract reported that more than 30 different amino acid derivatives were prepared in a reproducible manner.
2a. Shlyapnikov et al. [40,41]	The amino acids were added to the alcohol (ethyl, propyl, or isopropyl), and HCl or HBr was bubbled through until the amino acids completely dissolved. The solution was then heated at 60 to 80°C for 30 min. The resulting esters were reacted with acetic anhydride for 15 min at room temperature. Several different stationary phases and solid supports were tried. The ethyl derivatives were best separated with 0.5% and 3.0% PEG on Chromosorb W, the propyl derivatives with 10% PEGA on Celite 545, and the isopropyl derivatives with 1.5% PEG on Chromosorb W. All columns were operated isothermally between 120 and 190°C with a carrier flow adjusted to a constant value between 67 and 88 ml/min argon.	Volatile derivatives of the simple neutral and acidic amino acids were prepared. An attempt was made to find a polar stationary phase that would separate the N-acetylamino acid ethyl, propyl, or isopropyl esters in one operation. Successfully separated the N-acetylamino acid esters of ala, val, leu, ileu, gly, and pro under one particular set of conditions and the ester derivatives of met, ser, thr, asp, phe, and glu under a different set of conditions.

II. n-Amyl
A. N-Acetyl
1a. Johnson et al. [23]

The amino acids were suspended in 1-pentanol and the mixture was saturated with HBr within 2 to 4 min. The solution was then heated for 30 min at 165°C. The resulting esters were treated with acetic anhydride for 5 min at 26°C.

Glass columns 2 to 8 ft long packed with Carbowax 1540 or 6000 on Chromosorb W were used. The columns were maintained between 125 and 155°C. The injection port was heated to 300°C. The flow rate was adjusted to a constant value between 60 and 240 ml/min argon.

The N-acetyl -isobutyl, -isoamyl, and -n-butyl esters of ala, val, ileu, leu, gly, β-ala, and pro were also prepared. However, only a mixture of these amino acids as the N-acetyl-n-amyl esters could be separated under the conditions employed. Derivatives of 33 amino acids including all of the protein amino acids except try, cyS, arg, and his were prepared in the form of the N-acetyl-n-amyl esters. Preliminary studies on some derivatives indicate a yield of 85% or higher.

B. N-Trifluoroacetyl
1b. Blau and Darbre [5,6,9,10,11]

The amino acids were added to CH_3OH, and HCl was bubbled through for 20 min at 70°C. The amino acid esters were mixed with TFA anhydride at room temperature for 1 hr.

More than 50 different stationary liquid phases were tried under isothermal conditions. The usefulness of silicone-type, polyester-type, surfactants, metallo-organic, and several other stationary phases were tested.

Successfully prepared derivatives of most of the protein amino acids except try, his, arg, and cyS. It was determined that this procedure could be used for the quantitative estimation of almost all these derivatives using a gas density balance detector system.

A complete separation of all the prepared derivatives could not be accomplished on any one column. Most of the surface active agents appear to aid in the decomposition of the serine and threonine derivatives.

Table III. Butyl Ester Derivatives

Derivative	Preparation, column packing, and gas chromatography conditions	Results
I. n-Butyl		
A. Hydrochloride salt		
1a. Saroff et al. [38]	No information was given regarding the preparation of the ester hydrochlorides. A short 6 by .25 in. PEGA column was used for analysis of the individual ester hydrochlorides; the column was lengthened to 6 ft to separate mixtures. The column was operated isothermally at 131°C with a flow of 50 ml/min N_2.	The n-butyl ester hydrochloride salts were injected directly onto the column, and anhydrous NH_3 was added to the carrier gas to dissociate the salt and produce the free amino acid esters. This technique was successfully applied to the esters of val, gly, ileu, pro, asp, thr, met, ser, glu, phe, lys, and hypro. A mixture of six derivatives was very poorly resolved.
B. N-Acetyl		
1b. Youngs [54]	The amino acids were suspended in butyl alcohol, and the solution was saturated with HCl; the reaction was allowed to proceed for 45 min. The esters were treated with acetic anhydride for 1 hr at room temperature. The column consisted of a 6 ft by .25 in. copper tubing packed with 25% hydrogenated vegetable oil on firebrick; it was operated isothermally at 220°C with a flow rate of 80 ml/min helium.	A gelatin hydrolyzate was successfully analyzed for ala, gly, val, leu, ileu, and pro, although the leu and ileu derivatives did not separate.

2b. Shlyapnikov and Karpeiskii [[46]]

The amino acids were suspended in alcohol, and HCl was bubbled through until the amino acids went into solution. The reaction was allowed to proceed for 30 min at 60 to 80°C. The ester was then treated for 15 min with acetic anhydride.

Several different stationary phases were investigated. The best results were observed using 0.5% PEG on Chromosorb W. The column was operated isothermally between 165 and 170°C, and the carrier flow was adjusted to a constant value between 67 and 86 ml/min.

A mixture containing the derivatives of thr, ser, asp, met, phe, and glu and another containing ala, val, ileu, leu, gly, and pro were successfully separated using different GC conditions. The presence of alloisoleucine in the hydrolyzate of natural sporides-molide II, a cyclic depsipeptide, was determined using this procedure.

C. N-Trifluoroacetyl
1c. Zomzely et al. [[57]]

The amino acids were dissolved in 1-butanol containing 5% HCl, dimethylformamide, and dibutoxypropane. The solution was heated for 3 hr at 55 to 60°C. It was then neutralized with Na_2CO_3 and extracted with CH_2Cl_2. The excess CH_2Cl_2 was evaporated; the esters were redissolved in CH_2Cl_2 containing dimethyl formamide and TFA anhydride and reacted for 30 min at 28°C.

The 2 m by 0.63 cm i.d. stainless steel columns were packed with 1% NPGS on Gas Chrom A. The temperature was programmed from 75 to 220°C, the N_2 flow was 128 ml/min, and the injection block temperature was 265°C.

All the common protein amino acids including arg, his, and cyS were prepared. Only the asp and phe derivatives out of a mixture of 19 did not resolve during a single GC operation.

Table III. Continued

Derivative	Preparation, column packing, and gas chromatography conditions	Results
2c. Marcucci et al. [31]	The amino acids were esterified with 1-butanol containing dry Dowex 50 W × 4 cation exchange resin in the hydrogen ion form. The reaction was allowed to proceed for 3 hr at 130°C. The amino acid esters were removed from the resin by washing with citrate buffer. The esters were extracted from the wash with CH_2Cl_2 and then acylated with TFA anhydride. The 200 by 0.2 cm i.d. column was packed with 1% Carbowax 20 M on Chromosorb. The carrier flow was 14 ml/min N_2, and the temperature was programmed at 2.5°C/min from 90 to 200°C.	The derivatives of ala, val, ileu, leu, gly, pro, thr, met, asp, phe, and glu were prepared, and a mixture of these derivatives was successfully resolved. A 98% conversion to the ester was reported.
3c. Gehrke et al. [12,13,14,25,43]	The amino acids were dissolved in methanol containing 1.20 meq/ml HCl and reacted for 30 min at room temperature. The methyl esters were then interesterified with 1-butanol containing 1.20 meq/ml HCl for 3 hr at 90°C. The amino acid butyl esters were dissolved in CH_2Cl_2 and reacted with TFA anhydride for 2 hr at room temperature.	A single volatile derivative was obtained for each of the protein amino acids except try, which gave two derivatives, and arg, which gave none. Single volatile derivatives for try and arg could be prepared if the acylation was carried out in a sealed tube for 5 min at 150°C. Yields of 97% or better were reported for almost all the derivatives. Using the

3c. Continued

A mixed stationary phase containing 0.75% DEGS and 0.25% EGSS-X on Chromosorb W packed in a 100 × 0.4 cm i.d. glass column was used to resolve the derivatives. The temperature was programmed at 3.3°C/min from 67 to 218°C, and the N_2 carrier flow was 38 ml/min.

A mixed stationary phase, 21 amino acid derivatives could be separated and quantitated within an hour.

The amino acid residues obtained from the acid hydrolysis of gluten, gliaden, bovine serum albumin, and kappa casein were successfully separated and identified.

II. sec-Butyl
A. N-Trifluoroacetyl
1a. Pollock et al. [35]

The amino acids were added to sec-butyl alcohol in an ice bath; HCl was bubbled through the solution for 30 to 60 min, and it was then refluxed for 2 hr. Acylation of the amino acid esters was accomplished with methyl trifluoroacetate at room temperature for 2 to 24 hr.

A 300 ft by 0.01 in. i.d. UCON LB-500-X capillary column was used. It was operated isothermally at 140°C with a helium flow rate of 2 ml/min.

Successfully resolved 10 individual racemic amino acids into their antipodal N-TFA-sec-butyl esters. A mixture of 5 D,L amino acid derivatives were also separated using this procedure.

2a. Gil-Av et al. [15,16]

The amino acids were suspended in 2-butanol and bubbled with HCl for 7 hr at 100°C. The ester hydrochlorides were then dissolved in CH_2Cl_2 and reacted with TFA anhydride for 1 hr at room temperature.

A 150 ft by 0.01 in. i.d. polypropylene glycol LB-550-X capillary column was used. The helium carrier gas was set at 20 psi, and the column was operated isothermally at 150°C. Trifluoropropylmethylpolysiloxane and butanediol succinate capillary columns were also used.

Using these conditions the diastereoisomeric N-TFA-amino acid esters of ala, val, leu, pro, phe, met, hypro, lys, asp, and glu, were separated. A mixture of the diastereoisomers of ala, val, leu, pro, and phe derivatives were resolved.

Several N-TFA-2-n-octyl amino acid esters were also prepared.

Table IV. Miscellaneous Volatile Derivatives of Amino Acids

Derivative	Preparation, column packing, and gas chromatography conditions	Results
A. Aldehyde		
1a. Bier and Teitelbaum [4]	The amino acids in a ninhydrin solution were sealed in a glass ampule containing CCl₄ and placed in a boiling water bath was 30 min. The ampule was cooled, centrifuged, and the lower organic phase containing the aldehydes separated. Silicone Dow-Corning 200 columns operated at 78°C were used for GC analysis.	The aldehydes of val, leu, ileu, norleu, met, and phe were prepared. This technique is restricted to amino acids that can be oxidized to give volatile aldehydes.
2a. Hunter et al. [21]	The aldehydes that were formed during ninhydrin oxidation were swept from the solution by a stream of N₂ and collected in a series of cold traps. A 10 ft column packed with silicone impregnated Celite was used for GC analysis. The column was operated at 69°C with the helium flow set at 23 ml/min.	Successfully prepared aldehydes of val, leu, ileu, and ala.
3a. Zlatkis et al. [55,56]	The amino acid–ninhydrin solution was injected directly into a continuous H₂ flowing system. The amino acids were converted to volatile aldehydes by reaction with ninhydrin on diatomaceous	This procedure is limited to amino acids that can be oxidized to give volatile aldehydes. Ala, val, leu, and ileu were identified in a casein hydrolyzate using this technique.

3a. Continued

earth at 130°C. The aldehydes were swept into a GC column and separated. As each aldehyde emerged it was hydrocracked in a microreactor to produce methane and water. The water was selectively removed and the methane passed over a thermal conductivity cell.

A 10 ft by .25 in. copper tubing packed with a 10% equal mix of ethylene and propylene carbonates on C-22 firebrick was used for separation of the aldehydes. The column was operated isothermally at 25°C and the carrier gas was set at 100 ml/min H_2.

B. Trimethylsilyl
1b. Ruhlmann and Giesecke [36]

The N-trimethylsilylamino acid trimethylsilyl esters were prepared by reacting the amino acids with trimethylchlorosilan.

The 280 cm long column was packed with 30% silicone oil 12500 on Sterchamol. The column was operated at 165°C.

A mixture of the derivatives of ala, gly, val, leu, ileu, glu, and phe were prepared and separated.

C. Phenylthiohydantion (PTH)
1c. Pisano et al. [34]

The PTH amino acids were purchased in the pure form. The PTH derivatives of asp and glu were converted to the PTH methyl esters with BF_3-CH_3OH prior to GC analysis.

A 6 ft by 4 mm i.d. column was packed with 1% QF-1 on an inert support; it was operated isothermally at 175, 200, or 255°C.

Satisfactory results were obtained with most PTH derivatives except for those of ser, thr, and the basic amino acids.

Table IV. Continued

Derivative	Preparation, column packing, and gas chromatography conditions	Results
D. Pyrolysis products 1d. Winter and Albro [53]	The amino acid or protein was pyrolyzed at 300°C for 3 min, and the volatile pyrolysis products were then swept into the GC column. A 6 ft by 4 mm i.d. column was packed with 15% Quadrol over 5% KOH on silanized Chromosorb P; it was operated isothermally at 70°C.	The technique is based on the findings that each amino acid or protein produces a unique amine profile that can be used for qualitative purposes. This method was successfully applied to 16 common protein amino acids as well as to several proteins. It was reported that bovine albumin, egg albumin, and hemoglobin gave unique and reproducible amine profiles.
E. N-TFA-Dipeptide Methyl Esters 1e. Weygand et al. [47, 49, 52]	The N-TFA-amino acid thiophenyl ester was reacted with the free amino acid in hot glacial acetic acid for 2 hr. The resulting N-TFA-dipeptide was esterified with diazomethane. The 2 m column was packed with silicone grease impregnated Celite; it was operated isothermally above 190°C, and the helium flow rate was adjusted to a constant value between 91 and 100 ml/min.	These articles were concerned with the preparation of the N-TFA-amino acid and N-TFA-dipeptides and their conversion to the methyl ester. One investigation reported the separation of diastereoisomers of 34 different N-TFA-dipeptides methyl esters using a capillary column of the Golay type. Successfully separated some of the N-TFA-dipeptides obtained from a partial acid hydrolyzate of gelatin. The peptides were reacted with 0.2N HCl in CH_3OH and acylated with methyl TFA.

2e. Halpern and Westly [20]

The amino acid was esterified with a thionyl-methanol reagent. The methyl ester was then reacted with an excess of N-TFA-L-prolyl chloride.

The GC column was 5 ft long and packed with 5% SE-30 on Chromosorb W; it was operated either isothermally at 176°C. or programmed at 3°C/min from 176°C to 220°C with the N_2 flow rate at 28 ml/min.

The D,L isomers of ala, val, pro, met, and phe were separated as the N-TFA-L-prolyl dipeptide methyl esters.

a high-boiling derivative that is difficult to elute from many columns. Aspartic aoid, glutamic acid, methionine, phenylalanine, and proline can be easily prepared by many procedures.

Bayer et al. [1,2] and Nicholls et al. [33] have chromatographed several amino acid methyl esters as the free base. Negative results were obtained for cysteine, histidine, tyrosine, and tryptophan. The α-chloro-methyl ester derivatives are limited to simple neutral amino acids [32]. The α-hydroxymethyl esters were produced in low yields for several amino acids [45]. Losse et al. [29] had limited success with N-formyl-methyl esters. No suitable volatile dinitrophenyl-methyl esters were prepared for the hydroxy amino acids, histidine, tryptophan, and arginine [26, 27, 34]. The N-acetyl -ethyl, -propyl, and -isopropyl derivatives have been prepared for the simple neutral and acidic amino acids by Shlyapnikov et al. [39, 40]. Graff et al. [17] reported that they prepared 30-N-acetyl-n-propyl amino acid esters. Johnson et al. [23] prepared 33 N-acetyl-n-amyl derivatives, although negative results were obtained for tryptophan, cystine, arginine, and histidine. Ninhydrin oxidation is limited to amino acids that yield volatile aldehydes [4, 21, 55, 56]. Trimethylsilyl derivatives of the simple aliphatic amino acids, glutamic acid, and phenylalanine were prepared by Ruhlman and Giesecke [36]. Pisano et al. [34] obtained negative results for the phenylthiohydantoin derivatives of serine, threonine, and the basic amino acids.

Only the N-TFA -methyl* or -n-butyl derivatives have been reported to be suitable for GC analysis of the protein amino acids. Cruickshank and Sheehan [8] prepared the N-TFA-methyl esters of a mixture of 20 common amino acids within 50 min and obtained good reproducible yields for almost all derivatives. The procedure of Hagen and Black [19] required 4 hr to prepare a mixture of these same derivatives. They also determined that a constant proportion of each amino

---

*The following abbreviations will be used: DEGS (diethylene glycol succinate); EGSS-X (rhylene glycol succinate methyl silicone); NPA (neopentyl adipate); NPGS (neopentyl glycol succinate); NPS (neopentyl sebacate); PBDS (polybutanediol succinate); PEG (polyethylene glycol); PEGA (polyethylene glycol adipate); PhSi (phenyl silicone polymer); QF-1 (fluoroalkyl silicone polymer); SE-30 (methyl silicone); and TFA (trifluoroacetyl, trifluoroacetate, or trifluoroacetic).

acid was converted to the derivative. Makisumi and Saroff [30] presented extensive procedures for the synthesis of the individual N-TFA-methyl esters; they also presented melting and boiling point data for these derivatives. Their procedure required 2 hr to prepare volatile derivatives of amino acid mixtures. However, the reproducibility of their technique was not fully explored. Zomzely et al. [57] prepared the N-TFA-n-butyl esters of 19 naturally occurring amino acids within 4 hr. However, no quantitative data were presented. Gehrke et al. [12, 13, 14, 25] have presented the best quantitative study of volatile amino acid derivatives. They prepared the N-TFA-n-butyl derivatives of almost all the naturally occurring amino acids within 6 hr. The lowest yield obtained was 88% for cystine, while most of the remaining yields were 96% or better (see Table III, 3c). Using this procedure, a volatile derivative for arginine could not be prepared. However, Stalling and Gehrke [43] were able to trifluoroacetylate the n-butyl ester hydrochloride of arginine in 5 min at 150°C in a sealed tube. This gave 99% conversion to a single volatile derivative.

Darbre and Blau [10] reported that serious losses of N-TFA-alanine methyl, ethyl, n-propyl, and n-butyl esters occurred when subjected to a stream of 250 ml/min argon. No loss was observed with the N-TFA-n-amyl ester of alanine under the same conditions. Lamkin and Gehrke [25] reported that serious losses occurred while concentrating the N-TFA-methyl ester of valine that did not occur with the corresponding n-butyl derivative.

Methyl ester derivatives are the easiest esters to prepare. Conversion of the amino acid to the methyl ester is facilitated by the solubility of the amino acids in acidified methanol. The ready availability of diazomethane, a strong esterifying agent, also encouraged the preparation of methyl esters. In order to avoid losses due to excessive volatility, several workers have prepared esters from the higher boiling alcohols. However, this creates other problems. The basic amino acids, lysine, arginine, and histidine, as well as cystine, are not soluble in butyl or pentyl alcohols. Zomzely et al. [57] added dimethylformamide to n-butanol to solubilize the basic amino

acids, but this introduces an extraction step and the esterification required 3 hr. Lamkin and Gehrke [25] solved this problem by first preparing the amino acid methyl esters that are
soluble in n-butanol. Unfortunately, the interesterification
step required 3 hr.

Many workers have prepared N-acetyl ester derivatives,
but no one has reported the preparation of suitable volatile
acetyl derivatives for all protein amino acids. Lamkin and
Gehrke [25] reported that the N-acetyl-methyl ester of lysine
was not eluted under their GC conditions and that the N-acetyl-
n-butyl ester of lysine tailed badly and did not emerge until
216°C whereas the N-TFA-n-butyl ester emerged at 173°C.
The N-TFA esters are more volatile than the corresponding
N-acetyl derivatives, making it possible to elute the higher
boiling derivatives of lysine, arginine, tyrosine, histidine,
tryptophan, and cystine at reasonable temperatures.

The stability of the derivative must also be seriously
considered.   Darbre and Blau [11] examined the stability of
the TFA-n-amyl esters of cysteine, hydroxyproline, serine,
threonine, and tyrosine.   They found that these derivatives
were stable for long periods in carefully dried methyl ethyl
ketone or nitromethane but that progressive breakdown occurred in benzene, ethylenedichloride, n-amyl-TFA, or in
5% water in methyl ethyl ketone. Lamkin and Gehrke [25] reported that the n-butyl-TFA derivatives were stable for long
periods if carefully stored in anhydrous chloroform containing
a little TFA anhydride.

STATIONARY LIQUID PHASES AND GAS
CHROMATOGRAPHY CONDITIONS

The stationary liquid phase must be able to separate all
19 amino acid derivatives in order to be of maximum use for
most workers in the biological fields. It must also be stable,
even after repeated use, and give reproducible results. One
must also consider the column dimensions, carrier flow rate,
solid support, ratio of liquid phase to solid support, and tem-

perature conditions, that is, isothermal or temperature programmed.

The best separation of a complex mixture of derivatives was attained using temperature programming. However, Makisumi and Saroff [30] reported they have designed an instrument that will allow separation of derivatives of 21 common amino acids in $\frac{1}{2}$ hr under isothermal conditions. They constructed a GC instrument containing three different ovens; each operated isothermally at a different temperature. The sample was injected onto the column set at the highest temperature; by means of a timed valve switching operation the low and medium boiling components passed into the second column before any high boiling components were eluted from the first column. The medium and low boiling components were separated on a second column by a similar operation. The three groups of amino acid derivatives were then simultaneously separated into their individual components.

Darbre and Blau [10] tested over 50 different stationary phases. Unfortunately, the columns were operated isothermally and complete separation of the seven simple and hydroxy aliphatic amino acids as the N-TFA-n-amyl esters was not attained. Shlyapnikov and Kapeiskii [40], using isothermal conditions, were able to separate the N-acetyl -methyl, -ethyl, -propyl, and -butyl esters of the simple aliphatic amino acids and proline using 0.5% PEG on Chromosorb W. They also separated the N-acetyl -ethyl and -n-propyl esters of threonine, serine, aspartic acid, methionine, phenylalanine, and glutamic acid using 3% PEGA on Celite 545; the corresponding n-butyl derivatives were separated on 0.5% PEG on Chromosorb W. Blau and Darbre [6], using isothermal conditions, were able to separate the N-TFA-n-amyl esters of cysteine, hydroxyprolone, methionine, phenylalanine, aspartic acid, and glutamic acid with 5% MS 710 silicone on Silicel C22. Gehrke and Shahrokhi [13] determined that a mixed stationary phase of 0.75% DEGS and 0.25% EGSS-X on Chromosorb W could best resolve a 20-component mixture of n-butyl-N-TFA esters. A good separation of complex mixtures of N-TFA-methyl or N-TFA-n-butyl derivatives was achieved by Cruickshank and

Sheehan [8], Lamkin and Gehrke [25], Zomzely et al. [57], and Makisumi and Saroff [30] using NPGS coated supports, by Ikekawa [22] using a series column of NPGS followed by SE-30 on Gas Chrom P, by Hagen and Black [19] using Carbowax 20 M on Diataport S, and by Shlyapnikov et al. [42] using PBDS on Chromosorb W. Johnson et al. [23] achieved good separation of many N-acetyl-n-amyl esters with Carbowax 1540 or 6000 coated supports.

Klingman and McBride [24] found that 0.75% DEGS/0.25% EGSS-X mixed phase either on Chromosorb W or Gas Chrom P bleed excessively at temperatures over 160°C. In addition, the N-TFA-n-butyl derivatives of aspartic acid and phenyl-alanine did not separate using this column packing. The GC conditions were the same as those reported by Gehrke and Lamkin (see Table III, 3c), except that the program rate was 2.5°C/min, a single column system was employed, and a Microtek GC-2500 instrument, modified for on-column injection, was used. The mixed stationary phase could be stabilized by addition of a third factor (packing prepared by Applied Science Laboratories), but the aspartic acid and methionine peaks did not resolve, and the glycine-serine and isoleucine-threonine peaks did not completely separate.

A dual column system can be used to compensate for column bleeding and to stabilize the baseline drift that normally occurs when the temperature is programmed over a wide range.

Lamkin and Gehrke [25] observed that thermal decomposition of N-TFA-n-butyl amino acid esters occurred in heated metal injection ports, which was not observed with the N-acetyl butyl esters. Therefore, an all-glass system is recommended for TFA derivatives because decomposition could occur if metal columns are used. Direct on-column injection eliminates the effects due to metal ports.

DETECTORS

Thermal conductivity, gas-density balance, electron capture, argon ionization, and flame ionization detectors have all been used for GC analysis of amino acid derivatives. Thermal conductivity and gas-density balance detectors are not in wide use because they are not as sensitive as the ionization detectors and cannot be used with temperature-programmed systems. Landowne and Lipsky [26, 27] favored the use of the electron capture detector because of its high sensitivity (3 · 10^{-16} moles/sec), uniformity of response toward all DNP-amino acid methyl esters, and high selectivity. Electron capture is especially useful for detection of compounds containing halogen atoms and polar functional groups.

Argon and flame ionization detectors are most widely used. They have high sensitivity (10^{-9} moles and lower) for most volatile compounds. In addition, they can be used with temperature-programmed systems.

Darbre and Blau [9] reported a linear response for N-TFA-n-amyl esters using the gas-density balance detector. Landowne and Lipsky [27], using the electron capture detector, reported a linear response over the range 2-4 · 10^{-13} to 2-4 : 10^{-16} moles/sec for DNP-amino acid methyl esters. Hagen and Black [19] obtained a linear detector response between 1 and 20 μg amino acid as the N-TFA-methyl ester using flame ionization. Johnson et al. [23] reported a linear detector response from 10^{-7} to 10^{-10} moles of the N-acetylamino acid n-amyl ester using argon ionization.

APPLICATION

Gas chromatography is widely used for steroid, carbohydrate, and fatty acid analysis. Its use for analysis of amino acids is rapidly growing.

Several workers have analyzed the amino acid residues of different protein hydrolyzates using GC techniques. Bayer [1] identified several amino acid residues in an albumin hydrolyzate. The simple aliphatic amino acids were identified by Youngs [54] in a gelatin hydrolyzate and by Zlatkis et al. [56] and Melamed and Renard [32] in a casein hydrolyzate. The amino acid residues were identified by Cruickshank and Sheehan [8] in a ribonuclease hydrolyzate and by Gehrke and Shahrokhi [13] in gluten, gliaden, bovine serum albumin, and kappa casein hydrolyzates. Weygand et al. [50] separated some N-TFA-dipeptide methyl esters obtained from a partial acid hydrolysis of gelatin.

Gas chromatography has also been used to separate isomers of various amino acids. Shlaypnikov and Karpeiskii [40] ascertained the presence of alloisoleucine, instead of leucine, in the hydrolyzate of sporidesmolide II, a cyclic depsipeptide. Weygand et al. [51] succeeded in separating most of the diastereoisomers of 34 different N-TFA-dipeptide methyl esters. Pollock et al. [35] separated ten individual racemic amino acids into their antipodal N-TFA-sec-butyl esters. Halpern and Westly [20] used N-TFA-L-prolyl chloride to resolve D,L amino acids. Gil-Av et al. [15, 16] separated diastereoisomers of several N-TFA-2-n-butyl amino acid esters.

Klingman and McBride [24] analyzed the free amino acids obtained from human serum and rat nerve tissue using gas chromatography.

CONCLUSION

Only the N-TFA amino acid esters have been reported to give suitable volatile derivatives for all protein amino acids. Although the methyl esters are easier and quicker to prepare, the N-TFA butyl esters are preferred because they can be prepared without serious losses due to high volatility.

The preparation of n-butyl ester derivatives requires further investigation. The 3 hr normally required for n-butyl ester formation could be shortened if analogous reagents used in methyl ester preparation could be applied. The use of diazobutane or dibutylsulfite as esterifying agents could decrease the esterification period. A major drawback is finding a suitable solvent system to solubilize the basic amino acids.

It appears from the findings of Stalling and Gehrke [43] that their 2 hr acylation step could be shortened to a few minutes by performing the reaction at elevated temperatures in a sealed tube.

Temperature programming is necessary for complete separation of the 21 common amino acids. However, column bleeding occurs when the temperature is programmed over a wide range. This bleeding can be eliminated by using dual columns or by stabilizing the stationary liquid phase. The protein amino acids have been separated on NPGS and DEGS/EGSS-X coated supports.

Argon and flame ionization detectors are most widely used because of their versatility and high sensitivity.

The GC procedure of Gehrke and coworkers is more rapid and more sensitive than analysis carried out with the automatic amino acid analyzer. If six samples were prepared simultaneously according to the Gehrke procedure (see Table III, 3c), one could completely separate and quantitate six protein hydrolyzates within 12 hr. This is about one half of the time required to analyze six protein hydrolyzates using the amino acid analyzer. Furthermore, the gas chromatography instrument is more versatile because steroid, carbohydrate, fatty acid, and other biological organic acids can also be routinely, rapidly, and automatically separated and quantitated.

ACKNOWLEDGMENTS

Mr. McBride is a recipient of a Public Health Service Fellowship (No. 1-FM-GM-32, 889-01) from the National Institute of General Medical Sciences. This work was supported by grant NB-03697-05 from the National Institute of Neurological Disease and Blindness of the U. S. Public Health Service.

REFERENCES

1. E. Bayer, "Separation of Derivatives of Amino Acid Using Gas-Liquid Chromatography," in: D.H. Desty, ed., Gas Chromatography 1958, Academic Press, New York, 1958, pp. 333-339.
2. E. Bayer, K.H. Reuther, and F. Born, "Analysis of Amino Acid Mixtures by Gas Partition Chromatography" (in German), Angew. Chem. 69:640, 1957.
3. J.V. Benson and J.A. Patterson, "Improved Accelerated Automatic Analysis of Amino Acids," Fed. Proc. 23:371, 1964.
4. M. Bier and C. Teitelbaum, "Gas Chromatography in Amino Acid Analysis," Ann. N. Y. Acad. Sci. 72:641, 1963.

5. K. Blau and A. Darbre, "Preparation of Volatile Derivatives of Amino Acids for Gas Chromatography," Biochem. J. 88:8p, 1963.
6. K. Blau and A. Darbre, "Gas Chromatography of Volatile Amino Acid Derivatives. II. Leucine, Cysteine, Proline, Hydroxyproline, Methionine, Phenylanine, Aspartic Acid and Glutamic Acid," J. Chromatog. 17:445, 1965.
7. H. P. Burchfield and E. E. Storrs, Biochemical Application of Gas Chromatography, Academic Press, New York, 1962, pp. 573–588.
8. P.A. Cruickshank and J.C. Sheehan, "Gas Chromatographic Analysis of Amino Acids as N-Trifluoroacetylamino Acid Methyl Esters," Anal. Chem. 36:1191, 1964.
9. A. Darbre and K. Blau, "The Quantitative Estimation of Some Amino Acids by Gas Chromatography," Biochem. J. 88:8P, 1963.
10. A. Darbre and K. Blau, "Gas Chromatography of Volatile Amino Acid Derivatives. I. Alanine, Glycine, Valine, Leucine, Isoleucine, Serine, and Threonine," J. Chromatog. 17:31, 1965.
11. A. Darbre and K. Blau, "Trifluoroacetylated Amino Acid Esters: the Stability of the Derivatives of Cysteine, Hydroxyproline, Serine, Threonine, and Tyrosine," Biochem. Biophys. Acta. 100:298, 1965.
12. C.W. Gehrke, W.M. Lamkin, D.L. Stalling, and F. Shahrokhi, "Quantitative Gas Chromatography of Amino Acids," Biochem. Biophys. Res. Comm. 19:328, 1965.
13. C.W. Gehrke and F. Shahrokhi, "Chromatographic Separation of n-Butyl N-Trifluoroacetyl Esters of Amino Acids," Anal. Biochem. 15:97, 1966.
14. C.W. Gehrke and D.L. Stalling, "Acylation Studies on Arginine for Gas Chromatographic Analysis," Abstract Papers, 116c, No. 253, American Chemical Society Meeting, Atlantic City, New Jersey, 1965.
15. E. Gil-Av, R. Charles, and G. Fischer, "Resolution of Amino Acids by Gas Chromatography," J. Chromatog. 17:408, 1965.
16. E. Gil-Av, R. Charles-Sigler, G. Fischer, and D. Nurok, "Resolution of Optical Isomers by Gas-Liquid Partition Chromatography," J. Gas Chromatog. 4:51, 1966.
17. J. Graff, J.P. Wein, and M. Winitz, "Quantitative Determination of Alpha Acids by Gas-Liquid Chromatography," Fed. Proc. 22:244, 1963.
18. P. Hagen and W. Black, "A Method for the Quantitative Determination of the Composition of a Mixture of 19 Amino Acids by Gas Chromatography of Their N-Trifluoroacetyl Methyl Esters," Fed. Proc. 23:371, 1964.
19. P. Hagen and W. Black, "Gas Chromatographic Method for the Separation and Estimation of Amino Acid Derivatives," Can. J. Biochem. 43:309, 1965.
20. B. Halpern and J.W. Westly, "High Sensitivity Optical Resolution of D,L Amino Acids by Gas Chromatography," Biochem. Biophys. Res. Comm. 19:361, 1965.
21. I.R. Hunter, K.P. Dimmick, and J.W. Corse, "Determination of Amino Acids by Ninhydrin Oxidation and Gas Chromatography Separation of Leucine and Isoleucine," Chem. Ind. (London) p. 294, 1956.
22. N. Ikekawa, "Gas Chromatography of Amino Acids," J. Biochem. 54:279, 1963.
23. D.E. Johnson, S.J. Scott, and A. Meister, "Gas-Liquid Chromatography of Amino Acid Derivatives," Anal. Chem. 33:669, 1961.
24. J. Klingman and W. McBride, unpublished results.
25. W.M. Lamkin and C.W. Gehrke, "Quantitative Gas Chromatography of Amino Acids — Preparation of the n-Butyl N-Trifluoroacetyl Esters," Anal. Chem. 37:383, 1965.
26. R.A. Landowne and S.R. Lipsky, "Ultrasensitive Analysis of Amino Acids Via Gas Chromatography and Electron Capture Spectrometry," Fed. Proc. 22:235, 1963.

27. R.A. Landowne and S.R. Lipsky, "High-Sensitivity Detection of Amino Acids by Gas Chromatography and Electron Affinity Spectrometry," Nature 199:141, 1963.
28. A.B. Littlewood, Gas Chromatography, Academic Press, New York, 1962, pp. 446–448.
29. G. Losse, A. Losse, and J. Stock, "Gas Chromatography Separation of N-Formyl-Amino Acid Methyl Esters" (in German), Z. Naturforsch. 17b:785, 1962.
30. S. Makisumi and H.A. Saroff, "Preparation, Properties, and Gas Chromatography of the N-Trifluoroacetyl Esters of the Amino Acids," J. Gas Chromatog. 3:21, 1965.
31. F. Marcucci, E. Mussini, F. Poy, and P. Gagliardi, "Separation of Amino Acids and Their N-Trifluoroacetyl-n-butyl Esters by Gas Chromatography," J. Chromatog. 18:487, 1965.
32. N. Melamed and M. Renard, "Analysis of Mixtures of Amino Acids Using Gas Chromatography" (in French), J. Chromatog. 4:339, 1960.
33. C.H. Nicholls, S. Makisumi, and H.A. Saroff, "Gas Chromatography of the Methyl Esters of the Amino Acids as the Free Base and by Dissociation of Their Acid Salts," J. Chromatog. 11:327, 1963.
34. J.J. Pisano, W.J.A. Vandenheuvel, and E.C. Horning, "Gas Chromatography of Phenylthiohydantoin and Dinitrophenyl Derivatives of Amino Acids," Biochem. Biophys. Res. Comm. 7:82, 1962.
35. G.E. Pollock, V.l. Oyama, and R.D. Johnson, "Resolution of Racemic Amino Acids by Gas Chromatography," J. Gas Chromatog. 3:174, 1965.
36. K. Ruhlman and W. Giesecke, "Gas Chromatography of the Silylated Amino Acids" (in German), Angew. Chem. 73:113, 1961.
37. H.A. Saroff and A. Karmen, "Gas Chromatography of the N-Trifluoroacetyl-methyl Esters of the Amino Acids," Anal. Biochem. 1:344, 1960.
38. H.A. Saroff, A. Karmen, and J.W. Healy, "Gas Chromatography of the Amino Acid Esters in Ammonia," J. Chromatog. 9:122, 1962.
39. S.V. Shlyapnikov and M. Ya. Karpeiskii, "Comparative Study of Gas Chromatographic Separation of the Esters of N-Acyl Amino Acids," Biochem. (USSR) 29:1076, 1964.
40. S.V. Shlyapnikov and M. Ya. Karpeiskii, "Gas Chromatographic Separation of the Esters of N-Acetylamino Acids Using Polar Stationary Phases," Biochem. (USSR) 30:194, 1965.
41. S.V. Shlyapnikov, M. Ya. Karpeiskii, and E.F. Litvin, "The Use of Gas-Liquid Chromatography for Separation of Some Amino Acids," Biochem. (USSR) 28:544, 1963.
42. S.V. Shlyapnikov, M. Ya. Karpeiskii, L.M. Yakushina, and V.S. Oseledchik, "Application of Gas-Liquid Chromatography to the Quantitative Analysis of Some Amino Acids," Biochem. (USSR) 30:394, 1965.
43. D.L. Stalling and C.W. Gehrke, "Quantitative Analysis of Amino Acids by Gas Chromatography — Acylation of Arginine," Biochem. Biophys. Res. Comm. 22:329, 1966.
44. J. Wagner, "Separation and Determination of Amino Acids by Gas Chromatography" (in German), Angew. Chem. 72:588, 1960.
45. J. Wagner and G. Rausch, "Gas Chromatographic Separation and Determination of Amino Acids" (in German), Z. Anal. Chem. 194:350, 1963.
46. J. Wagner and G. Winkler, "Gas Chromatographic Separation and Determination of Amino Acids" (in German), Z. Anal. Chem. 183:1, 1961.
47. F. Weygand and R. Geiger, "N-Trifluroracetylation of Amino Acids in Anhydrous Trifluoroacetic Acid" (in German), Chem. Ber. 89:647, 1956.
48. F. Weygand and R. Geiger, "Separation of Amino Acids by Fractional Distillation of Their N-Trifluoroacetyl Methyl Esters" (in German), Chem. Ber. 92:2099, 1959.

49. F. Weygand, B. Kolb, and P. Kircher, "Gas Chromatographic Separation of the Methyl Esters of N-Trifluoroacetyl Dipeptides," Z. Anal. Chem. 181:396, 1961.
50. F. Weygand, B. Kolb, A. Prox, M. A. Tilak, and I. Tomida, "N-Trifluoro-acetylamino Acids. XIX. Gas Chromatographic Separation of N-TFA-Dipeptide Methyl Esters" (in German), Hoppe-Seyler's Z. Physiol. Chem. 322:38, 1960.
51. F. Weygand, A. Prox, L. Schmidhamer, and W. Konig, "Gas Chromatographic Investigation of Racemization During Peptide Synthesis," Angew. Chem. Internat. 2:183, 1963.
52. F. Weygand and A. Ropsch, "N-Trifluoroacetylamino Acids. XIV. N-Trifluoro-acetylations of Amino Acids and Peptides with Phenyl Trifluoroacetate" (in German), Chem. Ber. 92:2095, 1959.
53. L. N. Winter and P. W. Albro, "Differentiation of Amino Acids by Gas-Liquid Chromatography of Their Pyrolysis Products," J. Gas Chromatog. 2:1, 1964.
54. C. G. Youngs, "Analysis of Mixtures of Amino Acids by Gas Phase Chroma-tography," Anal. Chem. 31:1019, 1959.
55. A. Zlatkis and J. F. Oro, "Amino Acid Analysis by Reactor-Gas Chromatog-raphy," Anal. Chem. 30:1156, 1958.
56. A. Zlatkis, J. F. Oro, and A. P. Kimball, "Direct Amino Acid Analysis by Gas Chromatography," Anal. Chem. 32:162, 1960.
57. C. Zomzely, G. Marco, and E. Emery, "Gas Chromatography of the n-Butyl-N-Trifluoroacetyl Derivatives of Amino Acids," Anal. Chem. 34:1414, 1962.

Recent Advances in Applications of the Microcoulometric Titrating System

John A. Stamm

Dohrmann Instruments Company
Silver Spring, Maryland

Since the introduction of microcoulometric gas chromatography in 1960, considerable effort has been expended in the development of new fields of use for this unique analytical system. The instrumentation was used originally as a gas chromatographic detector and applied to the area of pesticide residue analysis because of the system's specific response to chlorinated and sulfur-containing organic compounds. In recent years, however, use of this versatile analyzer has been extended to include applications in the chemical, petroleum, natural gas, and steel industries. Current development effort has resulted in increasing the capability of the microcoulometric titrating system to include specific response to volatile nitrogen-containing compounds in addition to sulfur and halogen analysis. Continuing investigation indicates the possibility for specific carbon and phosphorus detection as well as selective response to certain classes of organic compounds.

All of these new developments involve electrochemical detection based on the unique concept of null-balance coulometry. As a word of brief explanation, the system functions by maintaining a constant titrant concentration in an electrolyte. Any material introduced into the electrolytic solution, which has a capability of reacting chemically with the titrant,

will result in a change in titrant concentration and an imbalance in the nulled system. This imbalance is immediately
sensed by the system, and titrant is generated from an electrode surface in order to maintain the null-balance condition
and return the titrant concentration to its original level. For
example, the titrant used in halogen determinations is silver
ion, and the imbalance in the nulled system results from pre-
cipitation of the insoluble silver halide. The number of cou-
lombs required for titration is displayed as a peak on a potentiometric strip-chart recorder; the area under the peak being
a measure of sample concentration. By judicious choice of
electrode and electrolyte compositions, it is possible to "tune"
the system to an element or class of compounds of particular
interest.

For most applications, the sample is passed into the detector as an effluent from a high-temperature furnace. The function of the furnace is to convert the sample to a physical
and/or chemical form suitable for electrochemical detection.
This is accomplished by either oxidation or reduction of the
sample, using oxygen or hydrogen as part of the carrier gas
stream. The only exceptions to this procedure are titration
of inorganic halides in aqueous media where the sample is
introduced as a liquid directly into the electrolyte and in
analysis of gaseous samples that will react chemically with
the titrant without any preoxidation or reduction. An example
of this would be the determination of odorants in natural gas,
where the volatile mercaptans react directly with silver ion to
form insoluble compounds.

As is the case with most instrumental methods of analysis,
sample handling techniques are of major concern. In this
regard, the microcoulometric titrating system is quite versatile. The unit readily accommodates the effluent from a gas
chromatograph where compositional analytical information and
specific detection are required. It is also a self-contained
analyzer for total assays that can accommodate solid, liquid,
and gas samples through the use of commercially available
sampling systems.

The most recent advance in technology, associated with

the microcoulometric titrating system, was the development of a titration cell that would give specific response to most nitrogen-containing compounds. The initial work in this area was done at the research laboratories of the Standard Oil Company of Indiana [1]. The new system is based on the principle of reducing the organic nitrogen compounds to ammonia in a stream of hydrogen over a bed of nickel turnings. The weak-base ammonia is titrated in a slightly acidic sodium sulfate electrolyte maintained at a constant pH. A pair of sensor electrodes senses any change in electrolyte pH, and a pair of generator electrodes generates hydrogen ions to restore the electrolyte to its former pH. With a lower detection limit of 1 nanogram (ng), the new system is capable of parts-per-million nitrogen analysis. In the case where acid-gas-forming elements, such as halogens, are present as possible interfering constituents, specificity of response is maintained by use of a scrubber (such as potassium carbonate) ahead of the titration cell. Typical applications of this system include determination of the nitrogen content of petroleum feed stocks as possible replacements for Kjehldahl and Dumas analyses, identification of nitrogen-containing fractions of petroleum products and crude oils, analysis of carbamate and other nitrogen-containing pesticides, and possible use for amino acid analysis.

Outside of the pesticide residue field, the petroleum industry has provided the largest area of application for the microcoulometric titrating system. To date, most of the interest has been in the area of sulfur analysis. Considerable work has been done using the system as a selective gas chromatographic detector to determine sulfur-compound distributions in various petroleum fractions. The pioneering investigations in this area were done by Esso in 1961 [2]. However, recent publications by the American Oil Company [3, 4] and the Universal Oil Products Company [5] have enlarged the scope of this work to a great degree. In this application, individual sulfur compounds may be determined in the presence of constituents that might otherwise interfere with chromatographic analysis utilizing nonspecific detectors. In addition, elution times of various classes of sulfur compounds may be compared directly to boiling-point data to characterize sulfur distributions of various petroleum fractions ranging from light gas oils and gasolines to heavy crudes.

Extending the range of this work to lower molecular-weight hydrocarbon species, the natural gas industry became interested in this system for the determination of odorants and other sulfur compounds present in natural gas streams. As early as 1960, the Institute of Gas Technology has been investigating techniques and equipment for this application [6]. It was felt that the most promising approach was to apply gas chromatographic separations, but this approach was limited because of hydrocarbon interference until the specific response and sensitivity of the microcoulometric titrating system was applied to the problem [7]. Results of preliminary investigations in this area were presented at the 1965 American Gas Association Conference [8]. As mentioned above, volatile mercaptans (used as odorants) will react with silver ion directly and may be determined without going through a combustion step. This suggests the possibility of screening for types of sulfur compounds using this discriminatory feature of the system.

In the previously described applications, use of the microcoulometric titrating system has been confined to monitoring the effluent from a gas chromatograph. In many instances, the need for compositional analytical information is minimal, and a rapid method for total elemental analysis is of prime concern. In these cases, the sample is introduced to the high-temperature furnace as a single slug, and the total halide, sulfur, or nitrogen content of the sample is measured by a single peak on a strip-chart recorder. Depending upon whether the sample is a solid, liquid, or gas, different approaches to sample handling will be required. As in the chromatographic application, a carrier gas is used to transport the sample in the vapor phase through the furnace and into the titration cell. Hydrogen or oxygen will be added, depending on whether the analysis is carried out under reductive or oxidative conditions. Gas samples may be introduced, using a gas syringe or a gas sample loop. Liquids may be injected with a microsyringe, utilizing a septum inlet similar to that used on a gas chromatograph. Solids create a little more of a sample-handling problem. However, furnaces are available where the sample may be weighed in a boat and introduced into the hot zone. In addition, there are solids-sampling systems that have recently become available commercially that appear to be satisfactory for use with the standard furnace design.

This "direct injection" application is where the microcoulometric titrating system has generated considerable interest over the past year. Examples of the types of analyses being performed using this technique follow:

1. Determination of dichlorobenzene in benzene.
2. Determination of thionaphthene in tetralin.
3. Determination of total chloride in polymers.
4. Determination of total sulfur in refinery streams.
5. Determination of total nitrogen in hydrocracking feed stocks.

There are many more such examples. However, these serve to indicate the broad spectrum of applications possible with this system. A paper, presented last August at the Pennsylvania State College Microchemical Symposium, discusses the determination of trace amounts of chloride in low-boiling organic liquids (kerosene range) and compares this method to procedures based on sodium biphenyl reagent, the ASTM wicklamp combustion, and modifications of tube combustion techniques [9].

The steel industry is using this system to monitor the effluent from combustion furnaces capable of handling solid iron and steel powders. In this case, interest concerns the determination of trace amounts of sulfur in metals. Use of the microcoulometric titration system to monitor the effluent from a DTA apparatus has also been investigated. Evolution of sulfur dioxide at specific temperatures gives information related to transition points of various metals and alloys. Papers presented at the 1966 Pittsburgh Analytical Conference describe this work in detail [10, 11].

It must be mentioned, in regard to this "direct injection" technique, that analysis times may now be measured in terms of minutes, rather than hours or days, when comparing to general "wet chemical" procedures. In addition, comparable precision and accuracy have been obtained, especially in applications where high sensitivity is a-prerequisite.

Investigations are being continued to extend the versatility of this unique analytical system even more. Preliminary work at Southwest Research Institute, published last year, indicates the possibility of specific response to organic phosphorus compounds [12]. Modifications of the electrode and electrolyte compositions suggest that specific response to carbon dioxide and organic compounds having unsaturated bonding might be obtained. The results of these recent advances related to the microcoulometric titrating system indicate that future application of electrochemical principles to the field of chemical analysis will provide the solution to some of today's challenging analytical problems.

REFERENCES

1. Ronald L. Martin, "Selective Detection System for Nitrogen Compounds," Seventeenth Pittsburgh Conference on Analytical Chemistry and Applied Spectroscopy, Pittsburgh, Pennsylvania, 1966.
2. P. J. Klaas, "Gas Chromatographic Determination of Sulfur Compounds in Naphthas Employing a Selective Detector," Anal. Chem. 33:1851, 1961.
3. R. L. Martin and J. A. Grant, "Determination of Sulfur-Compound Distributions in Petroleum Samples with Coulometric Detector," Anal. Chem. 37:644, 1965.
4. R. L. Martin and J. A. Grant, "Determination of Thiophenic Compounds by Types in Petroleum Samples," Anal. Chem. 37:649, 1965.
5. V. T. Brand and D. A. Keyworth, "A Rapid Determination of Individual Mercaptans in Gasoline Boiling Range Stocks," Anal. Chem. 37:1424, 1965.
6. B. H. Andreen and D. V. Kniebes, "Determination of Sulfur Compounds in Natural Gas by Chromatography," Paper CEP-62-13, American Gas Association Conference, Baltimore, Md., 1962.
7. E. M. Fredericks and G. A. Harlow, "Determination of Mercaptans in Sour Natural Gases by Gas Liquid Chromatography and Microcoulometric Titration," Anal. Chem. 36:263, 1964.
8. F. V. Wilby, "Determination of Sulfur Compounds in Natural Gas by Gas-Liquid Chromatography Using Low Temperature Sample Concentration, Temperature Programming, and Microcoulometric Detector," American Gas Association Production Conference, Buffalo, New York, 1965.

9. R. A. Hofstader, "Microdetermination of Trace Quantities of Organic Chloride by Combustion and Microcoulometry —Application to Low Boiling Liquid," Pennsylvania State Microchemical Symposium, State College, Pennsylvania, 1965.

10. W. R. Bandi, E. G. Buyok, and W. A. Straub, "The Determination of Trace Amounts of Sulfur in Iron and Steel," Seventeenth Pittsburgh Conference on Analytical Chemistry and Applied Spectroscopy, Pittsburgh, Pennsylvania, 1966.

11. W. Swartz and W. A. Straub, "Qualitative and Quantitative Study of Individual Metal Sulfides in Steels by a Combustion-Microcoulometric Method," Seventeenth Pittsburgh Conference on Analytical Chemistry and Applied Spectroscopy, Pittsburgh, Pennsylvania, 1966.

12. H. P. Burchfield, D. E. Johnson, J. W. Rhoades, and R. J. Wheeler, "Selective Detection of Phosphorus, Sulfur, and Halogen Compounds in the Gas Chromatography of Drugs and Pesticides," J. Gas Chromatog. 3:28, 1965.

Automatic Integrators and Gas Chromatography

Daniel M. Marmion

Allied Chemical Corporation
Industrial Chemicals Division
Buffalo Dye Plant
Buffalo, New York

It is generally desirable to simplify any technical operation. Simplification can mean such things as using less equipment, shortening the time involved, eliminating steps, or automating routine functions. Because of the number and complexity of samples analyzed in many gas-chromatography laboratories, simplification of the analysis is worthwhile. Automatic integration of peak areas can be one way to achieve this simplification.

The theory that explains the resolution of mixtures by gas chromatography has been the subject of many publications and is beyond the scope of this discussion. The same is true of the mechanics and electronics of integrators. What is pertinent, however, is how one gleans quantitative information from the chromatograms resulting from such separations. Quantitative gas chromatography generally requires a direct comparison between the peak areas of knowns and unknowns or a comparison between the area of an individual peak in a chromatogram and the sum of the areas of all the peaks in the chromatogram. The underlying principles of quantitation are quite simple. The trick, of course, is to make the necessary comparisons as quickly, cheaply, and accurately as possible.

Presently there are three popular ways of measuring peak areas: triangulation, planimetry, and automatic-mechanical or electrical integration. The method adopted is usually determined by the job to be done and by the money and manpower available. Each technique has its advantages and disadvantages; each should be considered as a possible way of doing the job.

For the purpose of this discussion, triangulation includes all techniques that require manually measuring peak height, width, or some combination or modification of these parameters, and subsequently relating these measurements to the quantity of the unknown. Triangulation, then, is in no way automatic and is of concern to us here only for purpose of comparison. Triangulation is slow and generally does not give the most accurate or precise results. But this method of integration has the advantage of requiring practically no equipment and demands very little from the analyst besides the ability to do simple mathematics.

In the broad sense, planimetry is a method of automatic integration, but for obvious reasons we will not consider it as such here. Instead, we shall consider it as a separate method of integration. Planimetry has the advantage that the necessary equipment is cheap. It also makes allowances to a great extent for some of the major sources of error in quantitative gas chromatography—asymmetric peaks and baseline drift. Results obtained with planimeters are usually more accurate than those obtained by triangulation, but considerable time and good judgment on the part of the analyst are needed to obtain them.

The third common means of measuring peak area, and one which is becoming increasingly popular, is the use of automatic or semiautomatic integrators. This method potentially offers the analyst the greatest simplification of quantitative gas chromatography. The advantages and disadvantages of the technique are the subjects of the following discussion.

Perhaps the most important advantage that integrators provide is direct saving of the analyst's time. Area determinations using techniques like planimetry or triangulation are time-consuming because the analyst must make several geometric measurements and calculations. On the other hand, integrators record areas directly in terms of trace readout or digital readout, either of which provides the necessary information in a relatively short period of time. Gas chromatography then becomes a more valuable tool for process control. This elimination of manual calculation also reduces the opportunity for human error and, as a bonus, releases the well-trained analyst from dull, repetitious work.

Both disc integrators and electronic integrators offer the analyst greater precision than triangulation. Experiments by many laboratories have shown that electronic integrators are 2 to 5 times more precise than ball and disc integrators and 4 to 10 times more precise than triangulation. Electronic integrators offer the best precision because they are not directly dependent on either recorder accuracy or human interpretive ability. This lack of recorder dependency is because electronic integrators receive their signal directly from the detector circuit and not from the recorder, as is the case with a ball and disc integrator. However, one should bear in mind that an overloaded or defective recorder in the system can cause electrical feedback, which in turn can affect the performance of the electronic integrator. Ball and disc integrators, of course, are by design directly dependent upon recorder accuracy and the operator's skill in interpreting the trace readout.

Wilkens Instrument and Research made an extensive evaluation of different methods of peak integration. Table I shows the results they reported at the 1965 meeting of ASTM Committee E-19.

Wilkens also compared digital and disc integrators when two types of columns were used to analyze a xylene mixture (Table II).

Tables I and II are similar to many tables in that they provide much useful information, but care should be taken in interpreting the data. First of all, small standard deviations should not be taken as proof that an integrator provides accurate results. Sales brochures frequently convey this impression. Such data testify only to the precision of a determination. They are a measure of how well a result can be repeated. They tell

Table I. Different Methods of Peak Integration

Technique	Standard deviation, %	Average applied time per chromatogram, min
Planimeter (expert operator)	4.06	45
Triangulation (drawing tangent to peak)	4.06	50
Peak height x width at one-half height	2.58	30
Cutting out and weighing a Xerox copy	1.74	120
Disc integrator:		
Operator 1	1.29	
Operator 2 (experienced)	0.85	
Operator 3	1.81	
Average of ten operators	0.89	
Digital integrator, tape readout	0.44	10

us nothing about the accuracy of the measurements. As H. W. Johnson, Jr., [8] states in his discussion on the performance of automatic integrators: "Testing a single integrating system by repeated GC analysis of the same sample is also indecisive. If results are poor, they can be attributed to the experimental part of the process. If results are uniform, this is a measure of precision, not accuracy."

One should also be cautious when considering the reported saving in analysis time. While the saving is real, it does not justify the purchase of such an expensive piece of equipment unless the samples analyzed are many and complicated.

Table II. Relative Standard Deviation*

	Packed bentone column	Capillary column
Digital integrator	0.13	0.28
Disc integrator	0.26	0.59

*in percent.

Electronic integrators offer several advantages not provided by disc integrators. They are able to bring gas chromatography much closer to automation by directly providing the analyst with area and retention time data. These data may be in the form of a visual or printed readout, which can be plugged immediately into computers, data loggers, punch cards, and so forth.   Gas chromatography then becomes a more practical and versatile on-stream control tool.

Because of the extremely high count rate of electronic integrators, the analyst can frequently elute samples at a very rapid rate and still obtain good quantitative results. The same rate of elution might confuse many strip-chart recorders, or it might produce peaks that are too difficult to measure by triangulation or with a disc integrator.

Since electronic integrators read directly off detectors or electrometer amplifiers, it is not always necessary to have a recorder in the system. However, it must be pointed out that recorders should be eliminated only when the nature of the sample being investigated is well known! A recorder tracing can sometimes be very revealing.

There is no doubt that the development of automatic integrators has been a blessing for gas chromatography, as they have done much to simplify the art.   But there are certain disadvantages, certain failings, connected with integrators that should be clearly understood before investing in such devices.

To start with, integrators are expensive.   The cost of the necessary equipment rises sharply from several hundreds of dollars for ball and disc integrators to several thousands of dollars for electronic integrators.   And this is usually only the beginning. Most integrators are complex, and their repair is beyond the ability of the average chromatographer. This means that the analyst must depend on plant instrument repairmen or factory servicemen to keep his equipment operating. Small laboratories frequently lack sufficient work to maintain an instrument shop, and factory service costs at least $100 a day and is not always readily available.

Mechanical and electrical integrators have greatly improved the precision of gas chromatography, but accuracy remains a problem and fluctuates with the type of device used.

Ball and disc integrators have relatively low count rates, and mechanical alterations are usually necessary to change this count rate. Low count rate means that numerous, sharp peaks confuse the instruments and reduce their accuracy. Accuracy is also affected by peak attenuation and the interpretative skill of the operator. Added to these difficulties is the fact that a ball and disc integrator can never be more accurate than the recorder from which it receives its signal.

In comparison, the accuracy of electronic integrators is certainly not limited by low count rate or recorder dependency. Modern instruments are capable of handling hundreds of thousands of counts per second and can operate independently of a recorder. Nevertheless, accuracy is far from guaranteed.

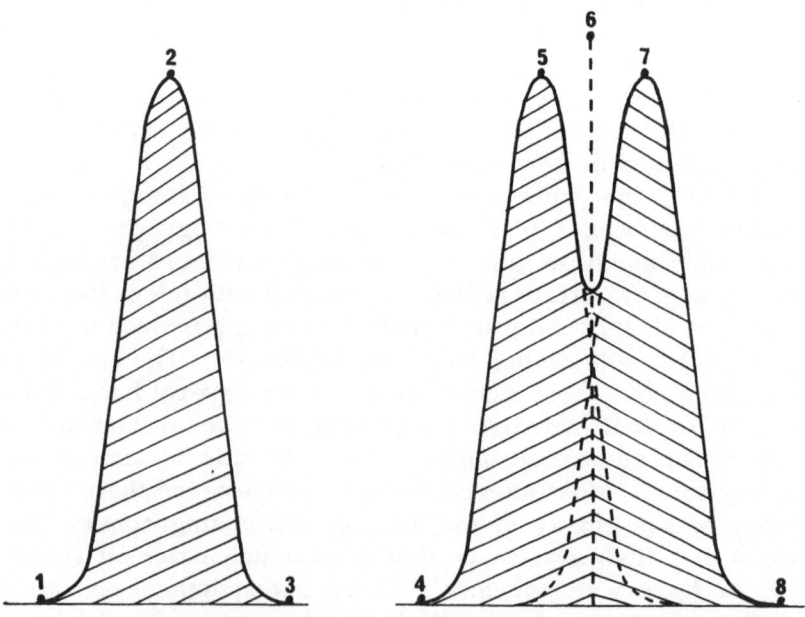

Fig. 1. In electronic integration counting normally begins when either the signal exceeds a certain triggering threshold (1 or 4) or when a valley between peaks is sensed (6). When a peak maximum has been reached (2, 5, or 7), retention time is recorded. Counting is then continued until either another between-peak valley is detected or until the signal drops below the triggering threshold and stays there for a predetermined length of time.

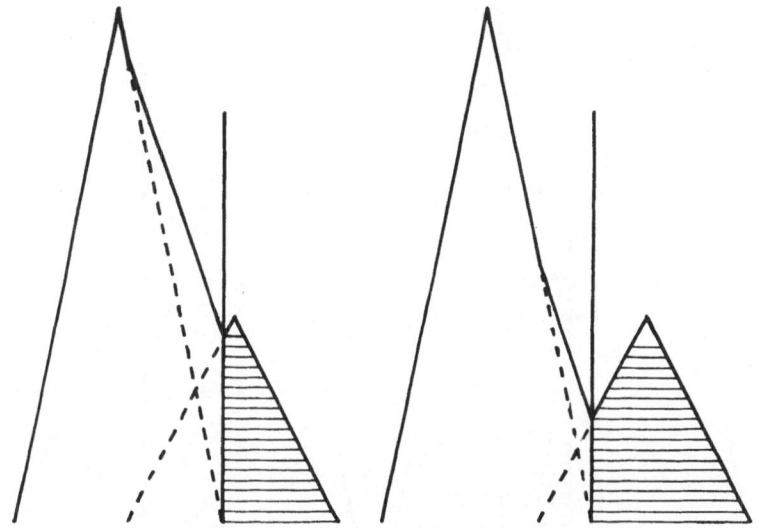

Fig. 2. Accuracy decreases as resolution decreases. If resolution is sufficiently poor, no interpeak valley is sensed, and both peaks are recorded as one.

The major causes of inaccuracy are poor resolution and drift in the chromatograph's baseline.

Electronic integrators are designed to count when either the chromatograph's signal exceeds a certain triggering threshold or the integrator senses a valley between peaks. Integration then continues until another between-peak valley is sensed or until the signal drops below the triggering value (see Fig. 1). The principle of operation sounds simple, but the integrator can be fooled. A chromatograph's detector and integrator can only sense the geometric sum of unresolved peaks; they cannot resolve the peaks. Consequently, two poorly resolved peaks are frequently recorded either as one peak or else two very inaccurate printouts result. Figures 2, 3, and 4 illustrate the effects poor resolution has on the accuracy of the areas measured. For simplicity, triangles have been used to represent the peaks usually obtained in gas chromatography. In each illustration the uppermost solid line represents the sum of the heights of the underlying triangles. The interpeak valleys sensed by the integrator are marked by

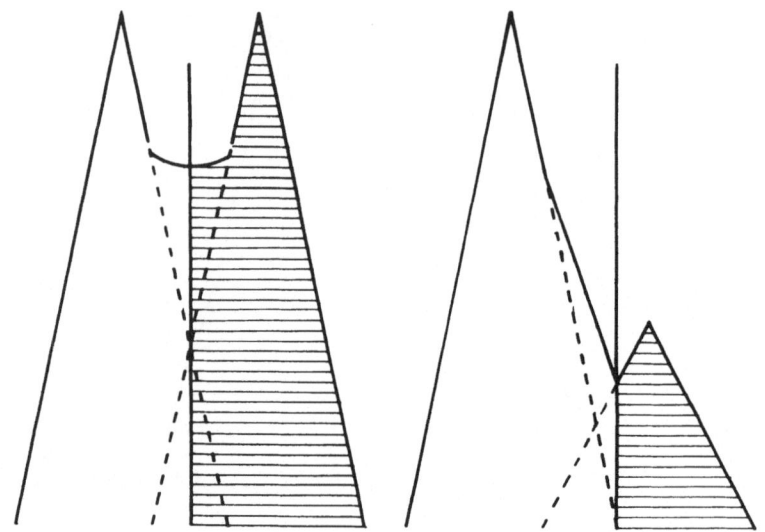

Fig. 3. The accuracy of area measurements of partially resolved peaks depends on the relative size of the peaks.

solid vertical lines. The areas recorded are those on either side of these lines.

As seen in Fig. 2, accuracy decreases as resolution de-creases. When resolution is sufficiently poor, no interpeak valley is sensed, and only one area is recorded for both peaks. Figure 3 demonstrates how the accuracy of area measure-ments depends on the relative sizes of the unresolved peaks. Good results are obtained for peaks that do not differ greatly in size. Poor results are obtained when one peak is but a small shoulder on another much larger peak. Figure 4 illustrates the dependence of accuracy on the symmetry of the peaks. Each peak in Fig. 4 is equal in area by construction. It is evident that asymmetric peaks lead to inaccurate area measurements. Some of the newer instruments feature circuits that can re-duce error due to poor resolution. But to make use of this equipment the analyst must first know the character of the curve being integrated. This means that he may have to run the sample twice — once to know what the corrections should be and once to apply the corrections. Running samples twice eliminates the time-saving feature of automatic integrators.

The shaded portion in Fig. 5a represents the error that might result from a drifting baseline. By the use of compensating equipment this error can be reduced to the smaller pie-shaped area shown in Fig. 5b. Obviously, the accuracy is reduced but not eliminated. The significance of any remaining error will depend upon the size of the peak being measured.

Besides baseline drift and poor resolution, inaccuracy can also result from a change in the response of the integrator's electronic components. This change can be a steady, gradual one due to the slow deterioration of certain parts, or it can be a fleeting change caused by "noise" in the chromatograph, in the integrator, or in the house electrical line. Anomalies such as these are difficult to detect and track down.

It should be apparent from the above discussion that integrators have many drawbacks. They are expensive to buy, expensive to maintain, and difficult to repair. And their accuracy is dependent upon the stability and efficiency of the chromatographic system as a whole. But integrators offer the analyst much in return: they save time and reduce human

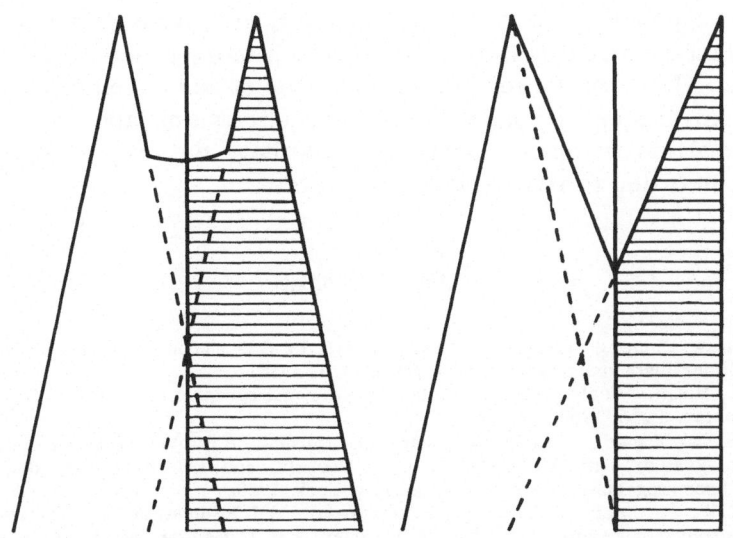

Fig. 4. Area measurements are less accurate when the unresolved peaks are asymmetric.

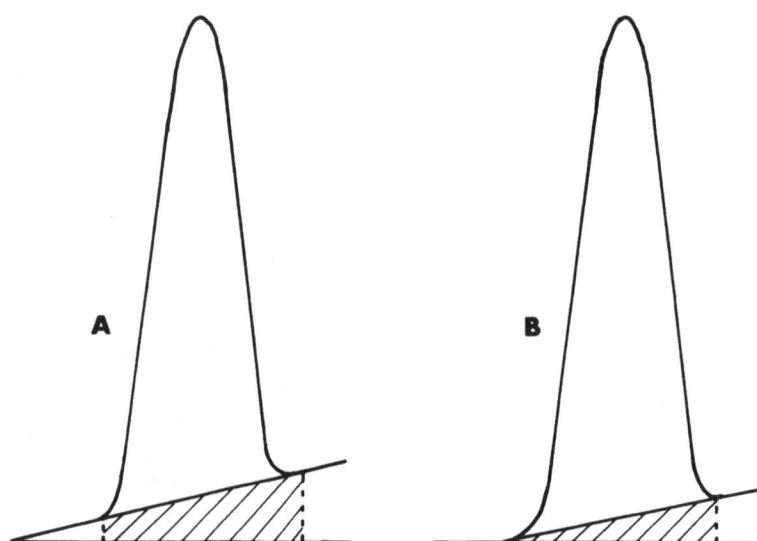

Fig. 5. Inaccuracy due to a drifting baseline can be reduced but rarely eliminated. A—without baseline drift compensation.   B—with baseline drift compensation.

error; provide more precise results; and, perhaps most importantly, make gas chromatography a more practical control tool. They are the key to the automation of gas chromatography and to processes dependent on gas chromatography. But as with all instrumental aids, one must have the proper application in order to realize these benefits.

REFERENCES

1. "New Printing Integrator Speed Analysis by Gas Chromatography," Perkin-Elmer Instrument News, Vol. 9, No. 4, p. 1, 1958.
2. Z. Bohm, "Analog Integrators for Gas-Chromatographic Apparatus," J. Chromatog. 3:265, 1960.
3. S. Dal Nogare, C.E. Bennett, and J.C. Harden, "A Simple Electromechanical Integrator," in: V.J. Coates, H.J. Noebels, and I.S. Fagerson, eds., Gas Chromatography, Academic Press, New York, 1958.
4. K.W. Gardiner, R.F. Klaver, F. Baumann, and J.F. Johnson, "Gas Chromatographic Chart Integrators," in: N. Brenner, J.E. Callen, and M.D. Weiss, eds., Gas Chromatography, Academic Press, New York, 1962.

5. I. Halasz and W.R. Marx, "Rapid Evaluation of Gas Chromatographic Analytical Data with the Aid of an Automatic Digital Integrator," Chem. Ing. Tech. 36:1115, 1964.
6. K.L. Jackson and C. Entenman, "A Two-Recorder Integrator Readout System for Gas-Liquid Chromatography," J. Chromatog. 4:435, 1960.
7. A.P.H. Jennings, "Recording Integrator for Gas Chromatography," J. Sci. Inst. 38:55, 1961.
8. H.W. Johnson, Jr., "Storage and Complete Computation of Chromatographic Data," Anal. Chem. 35:521, 1963.
9. R.D. Johnson, D.D. Lawson, and A.J. Havlik, "Voltage to Frequency Integrators in Gas Chromatography," J. Gas Chromatog. 3:303, 1965.
10. W.L. Perrine, "A Precision Integrator for Gas Chromatography," in: H.J. Noebels, R.F. Wall, and N. Brenner, eds., Gas Chromatography, Academic Press, New York, 1961.
11. D.T. Sawyer and J.K. Barr, "Evaluation of Several Integrators for Use in Gas Chromatography," Anal. Chem. 34:1213, 1962.
12. A. Strickler and W.S. Gallaway, "Analog Integration Techniques in Chromatographic Analysis," J. Chromatog. 5:185, 1961.
13. P. Van der Grinten and A. Dijkstra, "Integration with Ionization Detectors in Gas Chromatography," Nature 191, 1195:1961.

Measurement of Trace Amounts of Inert Gases in Blood by Gas Chromatography

Francis J. Klocke

Department of Medicine
State University of New York
Buffalo, New York

Although the physical characteristics of biologically inert*
gases make them uniquely suitable for a variety of biomedical
studies, limitations of methods for quantitating trace amounts
of the gases in blood, until recently, have made it feasible to
employ only those agents having commercially available
radioisotopes (e.g., Kr^{85} and Xe^{133}). Of particular interest are
gases that possess extremely low solubilities in blood, because
these gases tend to be excreted quantitatively during a single
passage through the lungs and recirculate negligibly when
introduced into the blood stream or a specific body tissue.
The use of gas chromatography for the analysis of trace
amounts of blood gases is a logical extension of the major
advances in chromatographic blood gas analysis that have
occurred in recent years and have been reviewed so excellently
in an earlier volume of this series by Albers and Farhi [1].
These authors covered methods for extraction of gases from
blood, injection of the extracted gases into the chromatograph
carrier gas stream, and specific analyses for CO_2, O_2, N_2, CO,
and several anesthetic and toxic gases. The present chapter

*In this chapter the term "inert" will be used in the biological sense to refer to
gases carried in the blood only in physical solution.

75

will not duplicate this material but instead will attempt to illustrate the factors involved in developing methods for the analyses of appreciably smaller concentrations of inert gases than those with which most previous workers have been concerned. The first portion of the chapter will consider several possible detector systems and the second a thermal conductivity unit that has been found adequate for the quantitative analysis of a fraction of a μl of an inert gas dissolved in a 2.0 ml blood sample. These will be followed by a summary of the limits of detection for the systems discussed and, finally, by a brief section on representative biomedical applications of such systems.

DETECTOR SYSTEMS

The present discussion is concerned with gases that traditionally have been measured with some form of thermal conductivity detector: H_2, He, Ne, N_2, A, SF_6, and so forth. In attempting to quantitate extremely small amounts of these gases, one must initially decide whether to continue with a thermal system or to switch to one of the nonthermal units recently developed for the inert gases.

Several of these newer systems are modifications of conventional ionization detectors. The coaxial microionization detector of Shahin and Lipsky [13] utilizes a tritium source, an applied potential of only 1 to 2 V, and an operating temperature of 150 to 200°C to effect substantial increases in the sensitivity of an argon detector to permanent gases. The electron drift-velocity detector of Smith and Fidiam [14] contains a tritium source near one electrode of a parallel plate argon ionization chamber and applies negative voltage pulses of short duration to this electrode to drive electrons from the ionized region toward the opposite electrode. The pulse duration is selected so that the appearance of sample gas in the argon carrier results in an increase of electron current to the collecting electrode.

The detectors of Bourke, Dawson, and Denton [4] and Hartmann and Dimick [7] employ helium rather than argon as a carrier, in order to take advantage of its appreciably higher ionization potential (19.8 eV as compared with 11.6 eV). The effective use of helium in this fashion requires an extremely high degree of helium purity, and both groups have invested considerable effort in this regard. The unit of Hartmann and Dimick also utilizes a particularly stable voltage source to operate at high field strengths and has been reported to have extremely low limits of detection.

The ultrasonic detector of Noble, Abel, and Cook [12] is noteworthy not only for its sensitivity but also for the clarity with which the theoretical considerations underlying its operation have been developed. Changes in the velocity of ultrasound produced by the presence of a sample gas in the carrier gas are detected by operating at a constant frequency and measuring the changes in wavelength accompanying the changes in velocity by determining the change in phase of the sine wave received by a transducer at one end of a gas-filled tube as compared with the phase of the sine wave transmitted from a second transducer at the other end of the tube.

Still another unit has been developed by Winefordner, Williams, and Miller [15]. Their detector cell is a modified capacitor and forms part of the tank circuit of a Clapp oscillator. The output of this oscillator is beat against a similar oscillator that contains a variable air capacitor for initial zeroing of the two oscillators to a zero beat frequency. The beat frequency is then measured with a frequency meter.

Although one or more of these units may ultimately prove the detector of choice for trace amounts of inert gases, some are still undergoing modification and, at this writing, none is available commercially in a well-established system. The limitation of most commercially available thermal conductivity systems, on the other hand, is that they are not capable of achieving the sensitivity required. For example, in one biomedical application to be mentioned below (the measurement of cardiac output), it is necessary to quantitate the hydrogen concentration of a 2.0 ml blood sample with an accuracy of 1 to 2% when the total amount of hydrogen in the sample is less than 0.0005 ml.

Thus, the possibility of utilizing a thermal conductivity detector resolves itself into the question of whether the signal to noise ratio of a conventional unit can be improved by one or two orders of magnitude. This, in turn, becomes the question of minimizing "baseline noise" caused by variations of detector current, carrier gas flow, and detector temperature. The latter is perhaps the most important of these factors and is traditionally minimized by housing the detector in a well-insulated air oven. Assuming that it is feasible to operate at temperatures below the boiling point of water, an alternative approach is to house the detector in a constant-temperature water bath, where temperature control to better than 0.01°C is readily achieved. A thermistor detector is especially suitable for such a bath because in contrast to a hot wire detector, it has its greatest sensitivity at relatively low temperatures and can be operated conveniently only a degree or two above ambient. Such temperatures are quite satisfactory for efficient separation of the inert gases under consideration and are easily maintained with a simple heater and stirring system. Thermistor detectors also have the advantage of being suitable for operation with carrier gases other than helium. Hot wire detectors have a limited "life expectancy" when operated at high currents in carrier gases of low thermal conductivity and, because they tend to be oxidized in the presence of oxygen, they are particularly unsuitable for operation with that gas as a carrier.

At least one thermal conductivity detector suitable for use in a water bath and possessing an appropriate sensitivity is available commercially (Model 1160 Standard Sealed Detector, Carle Instruments, Anaheim, California). The unit's electrical leads are sealed within a stainless steel conduit, and its small internal volume (0.058 ml) makes it capable of following separations effected by small diameter columns. Because of the relatively low currents required for satisfactory heating of the thermistor beads, the matching control unit (Model 1110 Micro Detector Control Unit) employs mercury reference cells as its source of power. The cells provide an extremely stable current supply for the usual Wheatstone bridge circuit and a "pure d-c" recorder input.

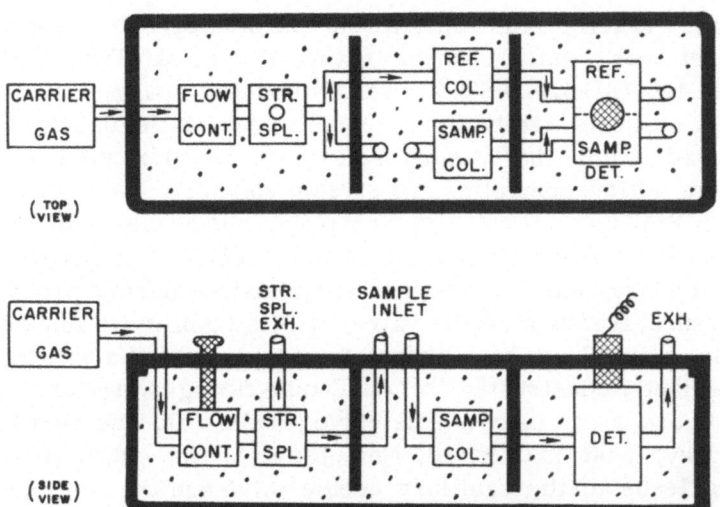

Fig. 1. Schematic drawing of high-sensitivity thermal conductivity gas chromatograph. (See text for details.) Flow cont = flow controller; str spl = stream splitter; str spl exh = stream splitter exhaust; ref col = reference column; samp col = sample column; det = detector (ref = reference side and samp = sample side); exh = exhaust.

HIGH-SENSITIVITY THERMAL CONDUCTIVITY GAS CHROMATOGRAPH

A gas chromatographic system incorporating the features just described is illustrated in Fig. 1. Except for the sample inlet, the entire unit is housed in a lucite water bath maintained at a constant temperature (usually 2 to 3°C above ambient) with a heater, mechanical stirrer, and proportional temperature controller (Model 72, Yellow Springs Instrument Company, Yellow Springs, Ohio). The chromatograph is constructed on a lucite frame which is suspended in the bath by four brass bolts and which can be easily removed for changing columns, etc. Research-grade carrier gas is supplied from a conventional gas cylinder equipped with a two-stage pressure regulator and short "clean-up column" of 13X molecular sieve. The latter serves to remove any trace of moisture (and perhaps other impurities) that may be present in the carrier cylinder.

The carrier gas enters the chromatograph through a bayonet-type, quick-connect fitting (Swagelok Tube Fitting Model 400-QC-61-DESO, Crawford Fitting Company, Cleveland, Ohio), which is easily disconnected whenever the unit is to be removed from the water bath. The carrier then passes through a flow controller and stream splitter. The controller (Model 8843 Extra-Low-Flow Controller with needle taper No. 1, Brooks Instrument Division, Emerson Electric Company, Hatfield, Pennsylvania) is designed to maintain a constant pressure difference across a needle valve, and its stability is improved by its being held at constant temperature. The splitter is conveniently constructed from a T-tube fitting (Swagelok Model 200-3) and a capillary needle valve (Model 151, The Matheson Company, East Rutherford, New Jersey). One outlet of the T fitting leads to the capillary needle valve and the other leads to the packed columns section of the chromatograph. The degree of stream splitting is readily adjusted with the screw control of the capillary needle valve. The carrier gas proceeding from the stream splitter to the packed columns section divides into two streams. The first leads to a sample inlet and then through an appropriate $\frac{1}{8}$ in. OD separating column to the sample side of the thermistor detector. The second leads through a needle valve (Brooks Model 8501) and/or a short $\frac{1}{8}$ in. OD "dummy" column to the reference side of the detector. The needle valve facilitates the adjustment of gas flow through the reference side of the detector, and the "dummy" column obviates the possibility of "baseline noise" related to an excessive void volume immediately upstream to the detector.

Sample gases are introduced into the carrier gas stream after having been extracted under vacuum in a modified volumetric Van Slyke apparatus and compressed in a small sampling loop, according to the technic of Farhi, Edwards, and Homma [6]. As discussed by Albers and Farhi [1], this type of sampling system provides a highly reproducible "slug" input and has the advantage of utilizing all the inert gas that is extracted from the blood sample. The tubing returning from the sampling loop to the sample inlet is filled with water vapor and carbon dioxide absorbents when the use of these substances is desirable [1]. The blood samples are originally collected in 2.0 ml precision glass syringes (Model 2YP, Becton, Dickinson, and Company, Rutherford, New Jersey), which are capped with mercury until the samples are introduced into the Van Slyke apparatus. The latter is accomplished by placing the glass tip of a syringe below a few milliliters of mercury in the Van Slyke cup, pressing gently against the bottom of the cup, and manipulating the main and leveling stopcocks in the usual fashion. Because the dead space at the tip of a syringe is only a few hundredths of a milliliter, one need not worry about remixing the blood within the syringe before introducing it into the Van Slyke apparatus unless the solubility of the gas being analyzed differs by a substantial amount in red blood cells and plasma. Excessive foaming of the blood during the extraction process is prevented by adding approximately 0.05 ml of gas-free capryl alcohol to the extraction chamber just before the blood sample is introduced. The latter is stored in the lower chamber of the volumetric Van Slyke apparatus [6].

It is particularly important in trace analyses of this kind to ensure that the gas sample enters the chromatograph in a reasonably small volume of carrier gas and that conditions within the chromatograph are arranged so that the gas of interest arrives at the detector in the maximum possible concentration. Thus, in addition to utilizing $\frac{1}{8}$ in. OD analytical columns, our procedure has been to employ a sampling loop constructed from capillary tubing and having a volume less than 1 ml and to minimize the length and diameter of tubing connecting the sampling loop to the sample inlet. The OD of the connecting tubing does not exceeed $\frac{1}{8}$ in. even when carbon dioxide and water vapor absorbents are employed. Carrier gas flows correspond to the relatively narrow columns and connecting tubing and are usually 2 to 8 ml/min for the sample and reference columns and approximately twice this amount for the stream splitter capillary needle valve. Most separations are effected with analytical column lengths of 4 to 12 ft.

The current at which the detector control unit is operated depends primarily on the carrier gas. In general, sensitivity is maximal when the detector thermistors are operated at currents slightly above those corresponding to the voltage maxima on their voltage–current curves [³]. Higher values are associated with reduced sensitivities and lower values with unmanageable instability and anomalous responses to sample gases. The voltage maximum of any given thermistor, of course, is dependent on several variables—among them ambient temperature and the thermal conductivity of the carrier gas to which the thermistor is exposed. Relatively low currents (e.g., a control unit setting of 6 to 10 mA) are optimal for gases such as oxygen or argon, while higher currents (e.g., a control unit setting of 12 to 20 mA) are required for gases such as helium or hydrogen. Because different sets of detector thermistors and the conditions of individual analyses can vary appreciably, it is often helpful to construct the voltage–current curves that obtain for a given detector in the actual analytical situation.

Perhaps it should also be mentioned that the system described can be constructed with rather simple shop facilities and for a total cost (excluding the strip-chart recorder) of less than $1500. If more than one unit is desired, additional savings can be effected by housing two or more chromatographs in a single water bath. The latter is particularly suitable when the time required for a single analysis makes it advantageous to have one individual operating two units simultaneously.

LIMITS OF DETECTION

The nonthermal conductivity detector systems listed above have all been reported to be one or two orders of magnitude more sensitive than conventional thermal conductivity units [4, 7, 12–15]. The same appears to be true of the water-immersed thermal conductivity unit. Detector sensitivities achieved for several gases with the latter are shown in Table I. Representative analytical records for the latter appear in Fig. 2.

On the basis of the discussion thus far, it does appear feasible to employ one or more of the detectors described with

Table I.   Limits of Detection Achieved
with High-Sensitivity Thermal
Conductivity Gas Chromatograph

Gas	g/sec	ml/sec	ml/2.0 ml blood*
H_2	$6 \cdot 10^{-12}$	$7 \cdot 10^{-8}$	$2 \cdot 10^{-6}$
He	$1 \cdot 10^{-11}$	$7 \cdot 10^{-8}$	$2 \cdot 10^{-6}$
CH_4	$3 \cdot 10^{-11}$	$4 \cdot 10^{-8}$	$1 \cdot 10^{-6}$
Ne	$3 \cdot 10^{-11}$	$3 \cdot 10^{-8}$	$1 \cdot 10^{-6}$
N_2	$5 \cdot 10^{-11}$	$4 \cdot 10^{-8}$	$1 \cdot 10^{-6}$
Ar	$4 \cdot 10^{-11}$	$2 \cdot 10^{-8}$	$8 \cdot 10^{-7}$
SF_6	$5 \cdot 10^{-11}$	$7 \cdot 10^{-9}$	$2 \cdot 10^{-7}$

*These limits were obtained using 2.0 ml blood samples
containing trace amounts of the gases listed. Each blood
sample was extracted in a Van Slyke apparatus etc., as
described in the text. The analytical conditions varied
for the different gases, although all gases except hydrogen
and helium were studied with a helium carrier gas.

an appropriate extraction, sampling, and separating system to
quantitate inert gases present in blood in concentrations below
0.0001 ml/ml.   Regardless of which system is used, however,
one must always contend with the problem of the other gases
present in the original blood sample. As pointed out by Farhi,
Edwards, and Homma [6], it is possible to extract from 1 ml
of arterial blood approximately 0.5 ml CO_2, 0.2 ml O_2, 0.01 ml
N_2, and 0.0002 ml Ar. Thus, if one is interested in measuring
0.0001 ml of an inert gas in 1 ml of blood, the inert gas may
comprise less than 1/1000 of the total amount of gas that can
be extracted from the blood sample.

This type of situation poses problems that can be handled
in various ways, all of which cannot be covered here. One
obvious example is the need for extreme care in ensuring that
the gas of interest passes through the detector only after
having been completely separated from all other gases (particularly CO_2 and O_2).   In the quantitation of blood hydrogen
concentration (the analysis with which the author has had the
greatest experience), CO_2 can be removed by including an
absorber such as ascarite (Arthur H. Thomas Company,

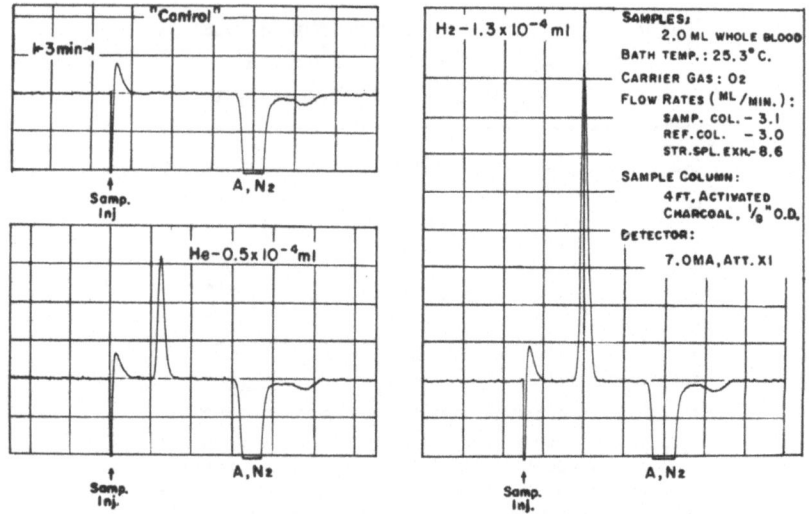

Fig. 2. Records illustrating the analysis of trace amounts of helium or hydrogen in blood. Analytical conditions are shown on the upper right. The record on the upper left is from a blood sample containing neither He nor H_2 and serves to verify that the blood contained no extractable substance that could have produced a peak interfering with those of He or H_2. The pressure disturbance associated with the injection of a sample is greater with an oxygen (or argon) carrier than with a helium carrier and is appreciably minimized by the flow control system described in the text. The large amount of O_2 extracted from the blood does not appear on the record because O_2 was employed as the carrier gas. CO_2 is not seen because a CO_2 absorbent was included in the tubing connecting the gas sampling loop to the chromatograph sample inlet. The peak height response of this system has been examined for He and H_2 concentrations up to 200 times those shown and deviates from perfect linearity over this entire range by no more than 4%.

Philadelphia) in the tubing connecting the sampling loop to the chromatograph sample inlet. Adverse effects of water vapor on the CO_2 absorber are eliminated by also including a water vapor absorber such as 3A molecular sieve upstream to the CO_2 absorber. The problem of O_2 can be obviated by employing O_2 as the carrier gas; although the thermal conductivity of H_2 differs from that of O_2 by a smaller amount than it does from that of Ar, the reduction in sensitivity resulting from the use of O_2 rather than Ar is small and easily tolerated. The problems of N_2 and Ar are quantitatively less important than those of CO_2 and O_2 and are readily handled with an analytical column of activated charcoal, 5A molecular sieve, or some other packed column adsorbent. The adsorbents (such as charcoal) that separate the H_2 from the N_2 and Ar without delaying the N_2 appreciably longer than the Ar are advantageous in that the total analytical time is less.

BIOMEDICAL APPLICATIONS

One current application of this type of analysis is the use of dissolved hydrogen as an indicator to measure cardiac output by the constant-rate-injection indicator–dilution technic [9]. Although a detailed treatment of indicator–dilution technics is not within the scope of this discussion, the method consists basically of infusing a solution of isotonic saline saturated with dissolved H_2 into the vascular system just upstream to the heart and then sampling the H_2 concentration of blood downstream to the heart after the blood has been mixed with the infusate but before it has passed through the lungs or body tissues. Cardiac output is then equal to the rate of infusion of saline times the ratio of the H_2 concentration of the original infusate to the H_2 concentration of the blood downstream to the heart. The major advantage of employing H_2 as an indicator is that its low solubility in blood (0.015 ml/ml-atm at 37°C) causes it to be virtually completely eliminated as the blood passes through the lungs, that is, unlike various dyes and other indicators, it recirculates negligibly. This fact makes the H_2 technic particularly suitable for repeated determinations of cardiac output over a short period of time and for studies of changes of cardiac output during an unsteady state. By the same token, the low solubility of H_2 also makes the ability to quantitate extremely low levels of H_2 concentration crucial to being able to maintain the infusion of H_2-saturated saline at an acceptably low level, for example, 0.5% of the total cardiac output.

One additional application of gas chromatographic analyses of blood H_2 concentration is to calibrate a "hydrogen electrode" in vivo. It has been known for several years that dissolved H_2

can produce a current by being oxidized at the surface of a platinum electrode [5] and, more recently, electrode circuits have been designed in which variations of current follow variations of dissolved H_2 concentration in a linear fashion [2, 8, 11]. Limitations in the usefulness of these electrodes have included "baseline drift," variations in response to any given hydrogen concentration over a short period of time, and sensitivity to several factors other than dissolved hydrogen. However, if blood samples are withdrawn while a hydrogen electrode is being used and the H_2 concentrations of the samples are determined chromatographically, it is possible to obtain an in vivo calibration of the electrode that greatly facilitates its being used quantitatively. This type of calibration has already been employed to measure the H_2 concentration of blood downstream to the heart in essentially continuous measurements of cardiac output by the constant-rate-injection technic. It also seems promising for several other cardiovascular studies.

ACKNOWLEDGMENT

The author wishes to acknowledge the assistance of Dr. Robert A. Klocke in designing and constructing the original model of the thermal conductivity gas chromatograph described above. This work has been reported in preliminary fashion elsewhere [10], and was supported by grants from the U.S. Public Health Service (HE 09587, NIH), the Junior Board of the Buffalo General Hospital, and the United Health Foundations, Inc.

REFERENCES

1. C. Albers and L.E. Farhi, "Analysis of Blood Gases by Gas Chromatography," in: L.R. Mattick and H.A. Szymanski, eds., Lectures on Gas Chromatography, 1964, Agricultural and Biological Applications, Plenum Press, New York, 1965, p. 163.
2. K. Aukland, B.F. Bower, and R.W. Berliner, "Measurement of Local Blood Flow with Hydrogen Gas," Circulation Res. 14:164, 1964.

3. J. A. Becker, C. B. Green, and G. L. Pearson, "Properties and Uses of Thermistors—Thermally Sensitive Resistors," Bell Sys. Tech. J. 76:170, 1947.

4. P. J. Bourke, R. W. Dawson, and W. H. Denton, "Detection of Volume Parts per Million of Permanent Gases in Helium," J. Chromatog. 14:387, 1964.

5. L. C. Clark, Jr., and L. M. Bargeron, Jr., "Left-to-Right Shunt Detection by an Intravascular Electrode with Hydrogen as an Indicator," Science 130:709, 1959.

6. L. E. Farhi, A. W. T. Edwards, and T. Homma, "Determination of Dissolved N_2 in Blood by Gas Chromatography and (a - A)N_2 Difference," J. Appl. Physiol. 18:97, 1963.

7. C. H. Hartmann and K. P. Dimick, "Helium Detector for Permanent Gases," J. Gas Chromatog. 4:163, 1966.

8. E. S. Hyman, "Linear System for Quantitating Hydrogen at a Platinum Electrode," Circulation Res. 9:1093, 1961.

9. F. J. Klocke and D. G. Greene, "Measurement of Cardiac Output with H_2," Fed. Proc. 25:205, 1966.

10. F. J. Klocke and R. A. Klocke, "Measurement of Trace Amounts of Inert Gases in Blood by Gas Chromatography," Physiologist 9:221, 1966.

11. W. A. Neely, M. D. Turner, J. D. Hardy, and W. D. Godfrey, "The Use of the Hydrogen Electrode to Measure Tissue Blood Flow," J. Surgical Res. 5:363, 1965.

12. F. W. Noble, K. Abel, and P. W. Cook, "Performance and Characteristics of an Ultrasonic Gas Chromatograph Effluent Detector," Anal. Chem. 36:1421, 1964.

13. M. M. Shahin and S. R. Lipsky, "The Mechanisms of Operation of a New and Highly Sensitive Ionization System for the Detection of Permanent Gases and Organic Vapors by Gas Chromatography," Anal. Chem. 35:467, 1963.

14. V. N. Smith and J. F. Fidiam, "Electron Drift-Velocity Detector for Gas Chromatography," Anal. Chem. 36:1739, 1964.

15. J. D. Winefordner, H. P. Williams, and C. D. Miller, "A High Sensitivity Detector for Gas Analysis," Anal. Chem. 37:161, 1965.

Gas Chromatography of Vitamin B_6 and Other Vitamins

Walter Korytnyk

Department of Experimental Therapeutics
Roswell Park Memorial Institute
Buffalo, New York

Although a number of methods are available for the assay of vitamins [1-5], gas chromatography, where applicable, may offer unique advantages in speed of separation, sensitivity, and convenience in quantitation. The recently introduced combination of gas chromatography with mass spectrometry [6] can be expected to provide a powerful tool for analytical studies in this field, such as those concerned with the metabolic fates of vitamins and their analogs. The present chapter embodies an attempt to indicate the potential usefulness of gas chromatography in vitamin research in general and in the study of vitamin B_6 in particular.

Vitamin B_6, which is widely distributed in nature, plays an essential part in amino acid metabolism [7] and also has important functions in the metabolism of carbohydrates [8] and possibly fats [9]. Because of the importance of this vitamin, considerable attention is being given to its mode of action and metabolism [10]. Antagonists of vitamin B_6 are being synthesized and tested in efforts to find selective growth inhibitors applicable in the chemotherapy of cancer and other diseases as well as to learn more about the properties of this vitamin and the enzymes involved in its function [11]. Work of this nature can be greatly facilitated, in certain instances, through the use of gas chromatography [12].

Vitamin B_6 is not just one compound, but a group of compounds: pyridoxol (I), pyridoxal (II), pyridoxamine (III), and their respective a^5 phosphates (IV, V, and VI). The name "pyridoxine," although originally limited to pyridoxol, is now being extended to all such substances with vitamin B_6 activity. Simple pyridine derivatives are amenable to gas chromatography, even using the conventional thick-film columns, but the presence of polar groups in the vitamin B_6 compounds tends to keep them from being volatile enough for this method of analysis. Several derivatives of pyridoxol are known, however, which are reasonably volatile. Thus a^4, 3-O-isopropylidene-pyridoxol (VII, G = CH_2OH), obtainable in quantitative yield from pyridoxol by HCl–catalyzed condensation with acetone [13], can be sublimed at 100° and 15 mm Hg. Likewise, tri-O-acetylpyridoxol (VIII), which has been obtained in high yield from pyridoxol, has been distilled at 85 to 90° and 10^{-4} mm Hg [14]. Under the circumstances, my associates and I initially studied the gas chromatography of a^4, 3-O-isopropylidene and tri-O-acetyl derivatives and later found it possible to extend our work to trimethylsilyl and 3-O-benzyl derivatives.

The conditions used for gas chromatography were similar for all derivatives studied in this investigation and were as follows.

Apparatus. An F & M Model 400 Biochemical Gas Chromatograph equipped with a hydrogen flame detector and a direct injection port was utilized. The column was a 4 ft × $\frac{1}{4}$ in. glass column packed with 1.2% silicone gum rubber SE–20 on 80- to 100-mesh Chromsorb G that had been washed with acid and silanized.

GLC Conditions. The column temperature was varied between 115 and 185°, depending on the individual sample, and the temperatures of the injection port and detector block were maintained 50° above the working column temperature. The flow rate was 75 ml/min (inlet pressure 30 psi) for helium, 50 ml/min for hydrogen, and 450 ml/min for air. A typical injection sample had a volume of 1 to 2 μl, and contained 3 to 4 μl of any single component.

ISOPROPYLIDENE DERIVATIVES

Isopropylidene (acetonide) derivatives of vitamin B_6 compounds can be obtained in high yield whenever the 3- and a^4-hydroxyl groups [13] (or sometimes the a^4- and a^5-hydroxyls [15]) are free. The isopropylidene derivatives subjected to gas chromatography gave sharp and symmetrical peaks, with the exception of two carboxylic acids (Table I, compounds 7 and 8),

Table I. Relative Retention Times of Pyridoxol Analogs

Nature of G	Computed No.	G	Relative retention time*	
			Column temp. 125°	Column temp. 150°
Alcoholic	1	CH_2OH	1.00	1.00
	2	$CHOHCH_3$	1.20	1.20
	3	$COH(CH_3)_2$	1.24	—
	4	CH_2CH_2OH	1.58	1.55
	5	$CH_2CH_2CH_2OH$	2.59	2.28
	6	$CH_2CH_2CH_2CH_2OH$	4.15	3.28
Carboxylic	7	$COOH$	1.42	1.55
	8	CH_2COOH	2.91	2.24
	9	CH_2CH_2COOH	—	3.42
	10	$CH_2CH_2CH_2COOH$	—	4.92
Alkylic	11	CH_3	0.374	—
	12	$CH_2CH_2CH_3$	0.734	—
Miscellaneous	13	NH_2	0.949	1.03
	14	CHO	0.480	—
	15	$COCH_3$	0.670	0.82
	16	CH_2CN	1.23	—
	17	$COOCH_3$	0.936	1.0
a^4, a^5-Isopropylidenepyridoxol			1.38	—

*Retention times are relative to a^4,3-O-isopropylidenepyridoxol ($G = CH_2OH$), which has retention times of 3.4 min at 125° and 1.0 min at 150°.

which gave some additional minor peaks indicative of de-
carboxylation. The retention times were determined at 125
and 150°, relative to a^4, 3-O-isopropylidenepyridoxol (VII,
$G = CH_2OH$; Table I). The isomeric cyclic ketal, a^4, a^5-isopro-
pylidenepyridoxol (IX), has a somewhat greater retention
time (Table I) than the reference compound.

The differences in the retention times appear to be
associated with volatility, which in turn is determined by
various factors, such as intermolecular hydrogen bonding and
Van der Waals interactions. For example, the lowest retention
times are characteristic of alkyl derivatives, followed by
alcohols and acids. Within any group of similar compounds,
the retention time increases regularly with the length of the
side chain. When the logarithm of the retention time is plotted
against the number of carbons in the side chain, a straight line
is obtained for the straight-chain homologs (Fig. 1); but
branching depresses the retention times, and they do not
conform to the straight-line relationship (compare compounds
2 and 3 in Table I). Branching apparently reduces Van der
Waals binding between the molecules in the gas and liquid
phases. The differences between the retention times at two
different temperatures (125 and 150°) increases regularly with
the length of the side chain for the alcohols (Fig. 1). Thus, the
gas chromatography column exhibits considerable selectivity
for different structural modifications in the 5 position. These
results illustrate how retention-time data can be used to obtain

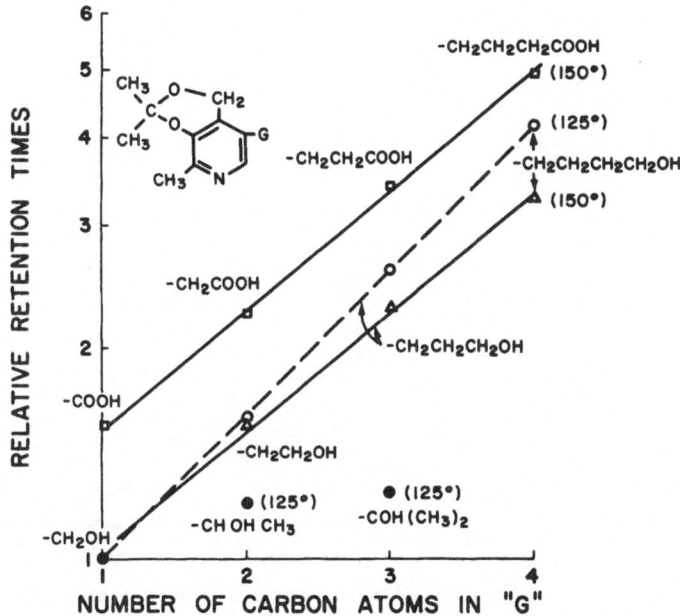

Fig. 1. Relationship between structures and relative retention times of isopropylidene derivatives with different substituents G.

structural information, such as the length of the side chain or the nature of the end group.

ACETYL DERIVATIVES

Although isopropylidene derivatives of vitamin B_6 compounds are usually suitable for gas chromatography, unfortunately the isopropylidene group cannot be introduced into all forms of the vitamin. A general procedure for estimating all forms of the vitamin as well as some of its metabolites and antimetabolites would require a generally applicable method for forming derivatives. Acetylation was studied in detail, but procedures for the synthesis and gas chromatography of other derivatives (e.g., trimethylsilyl and benzyl derivatives) have also been studied and are discussed later in this review.

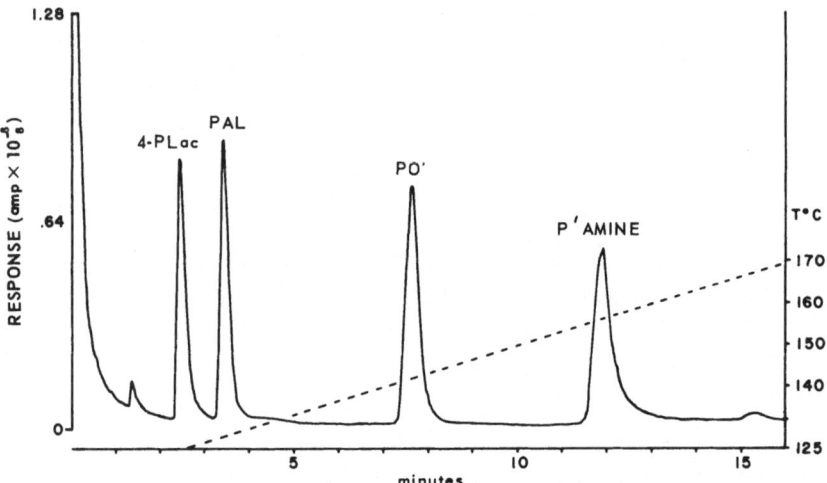

Fig. 2. Chromatogram of acetates of, 4-pyridoxic acid lactone (4-PLac), pyridoxal
(PAL), pyridoxol (POL), and pyridoxamine (P'AMINE).

Table II. Retention Time of Acetates of Pyridoxine
Compounds at 125°

Acetate of	Time, min	Time relative to pyridoxol acetate
Pyridoxol	8.5	1.00
Pyridoxamine*	19.6	2.31
Pyridoxal	3.1	0.36
Pyridoxal ethyl acetal	2.4	0.27
4-Pyridoxic acid lactone	2.2	0.26
5-Pyridoxic acid lactone	2.8	0.33
4-Deoxypyridoxol	3.0	0.35

*At 165°, the retention time was 2.3 min.

Peracetylated derivatives of the natural pyridoxine compounds have been described previously [14], with the exception of the acetate of pyridoxal. 3, a^4-O-Diacetylpyridoxal (X) was obtained by shaking pyridoxal hydrochloride with a mixture of pyridine and acetic anhydride overnight. Its hemiacetal structure (X) was confirmed by NMR spectroscopy, including analogy with the NMR spectrum of dipalmitoylpyridoxal (absence

H₃C, C, O, CH₂, G, H₃C, O, H₃C, N   VII

CH₂OAc, AcO, CH₂OAc, H₃C, N   VIII

CH₃, CH₂O—C, HO, CH₂O, CH₃, H₃C, N   IX

AcO, H, C, O, AcO, CH₂, H₃C, N   X

of aldehyde proton, 5-CH_2 proton appearing as an AB quadruplet) [16].

The retention times of the acetates are summarized in Table II. The retention data indicate that the three non-phosphorylated forms of the vitamin and the most important metabolite, 4-pyridoxic acid lactone, can be separated as their acetates. For separating the three vitamers and 4-pyridoxic acid lactone, it was convenient to use temperature programming as indicated in Fig. 2. In this way, separation of the four compounds can be accomplished within 13 min [17]. Acetylation was carried out by treatment or the hydrochlorides or free bases of the vitamers of their analogs with a mixture of acetic anhydride and pyridine (1:1 by volume) at room temperature for 2 to 4 hr. The acetylation mixture was then injected directly on the column.

Pyridoxal may interfere with the analysis of 4-deoxy-pyridoxol and of 5-pyridoxic acid lactone because of the similarity of the retention times (Table II). In such cases, it

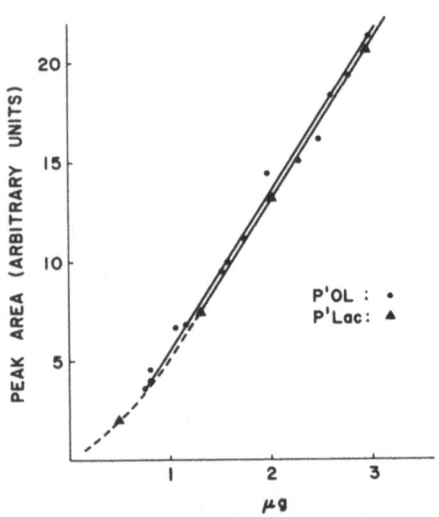

Fig. 3. Response in terms of peak areas to different amounts of acetates of pyridoxol (P'OL) and of 4-pyridoxic acid lactone (P'Lac).

Table III. Analysis of Synthetic Mixtures of Pyridoxol Acetate and 4-Pyridoxic Acid Lactone Acetate

Sample No.	Pyridoxol acetate			4-Pyridoxic acid lactone acetate		
	Calc. wt %	Calc. mol. %	Determined area %	Calc. wt %	Calc. mol. %	Determined area %
1	60.9	51.9	57.9	39.1	48.1	42.2
2	75.8	68.6	73.5	24.2	31.4	26.4
3	81.4	75.4	80.3	18.6	24.6	19.7

may be expedient to convert pyridoxal to its hemiacetal, which can be accomplished quite readily and quantitatively by refluxing the reaction mixture with ethanol for 2 hr prior to acetylation. Alternatively, pyridoxal may be reduced quantitatively to pyridoxol through the use of sodium borohydride.

The quantitative aspects of the acetylation procedure have been studied, and the response to various amounts of the acetyl derivatives of pyridoxol and 4-pyridoxic acid lactone has been determined (Fig. 3). There is a linear and almost equal response to both acetyl derivatives in the range of 0.8 to 15.0 μg. This was borne out by analysis of a synthetic mixture of these two acetyl derivatives (Table III). It is to be noted that the values found correspond more closely to the actual weights of the acetyl derivatives than to the mole percentages. Under the conditions used, pyridoxol was found to be acetylated quantitatively within experimental error. This was also the case with the acetyl derivatives of other vitamin B_6 compounds, as was demonstrated by the use of internal standards.

TRIMETHYLSILYL (TMS) DERIVATIVES

Trimethylsilylation of a number of natural products has made them amenable to gas chromatography. As applied to vitamin B_6 compounds, trimethylsilylation offers the promise of extending gas chromatography to the phosphorylated derivatives, in much the same way as recently has been done in the case of the ribonucleotides, by the direct trimethylsilylation of both the phosphate and the hydroxyl functions [18].

Gas chromatography of trimethylsilyl derivatives of non-phosphorylated pyridoxine compounds was performed at a column temperature of 115°. Both cyclic ketals of pyridoxol (VII, G = CH_2OH; IX) gave TMS derivatives under conditions that had been found suitable for carbohydrates [19]. The retention times were 6.9 min for a^4,3-O-isopropylidenepyridoxol TMS ether and 9.7 min for a^4,a^5-O-isopropylidenepyridoxol TMS ether. The conversion was quantitative for the ketal with the six-member ring, but not for the one with the seven-member ring. Pyridoxal gave rise to only a single peak, one with a retention time of 6.8 min, but formation of the TMS derivative of pyridoxol created problems. Generally, trimethylsilylation of pyridoxol gave rise to two peaks: brief exposure (several minutes) to the trimethylsilylation mixture gave a peak with a long retention time (17.0 min); but on longer standing, a peak with a short retention time predominated. Addition of dimethyl sulfoxide [20] to the reaction mixture increased the rate of formation of the peak with the short retention time; and after the mixture was left standing overnight, the peak with the long retention time had virtually disappeared. The latter peak is formed exclusively, however, if pyridoxol is taken up in dimethyl sulfoxide and is shaken with hexamethyldisalazane overnight. Similar reaction conditions are also conductive to the formation of the TMS derivative of pyridoxal.

Trimethylsilyl derivatives of phosphorylated vitamin B_6 compounds were subjecteu to gas chromatography at a column temperature of 175°. The a^5-phosphates of pyridoxol, pyridoxal, and deoxypyridoxol, after being subjected to trimethylsilylation under standard conditions, gave simple sharp peaks at retention times of 5.0, 6.9, and 3.4 min, respectively. Thus, most of the pyridoxine phosphates and analogs can be subjected to gas chromatography as their trimethylsilyl derivatives. Nevertheless, the quantitative aspects of the trimethylsilylation reaction as it applies to the gas chromatography of the vitamin B_6 compounds remain to be studied.

3-O-BENZYL DERIVATIVES

One convenient blocking group in the synthesis of vitamin B_6 analogs is the benzyl group. Benzylation of the phenolic hydroxyl in pyridoxol, pyridoxal, and similar compounds can be achieved directly by the interaction of these substances with dimethylphenylbenzylammonium hydroxide [21]. Most of the resulting derivatives can be subjected to gas chromatography. When the column temperature is kept at 150°, the retention times of some of these 3-O-benzylated compounds are as follows:

3-O-benzylpyridoxal, 7.0 min; 3-O-benzyl-4-pyridoxic acid lactone, 6.0 min; and 3-O-benzyl-5-pyridoxic acid lactone, 8.5 min.

POSSIBILITIES OF GAS CHROMATOGRAPHY IN VITAMIN B_6 RESEARCH

It has been shown that the various forms of vitamin B_6 can be subjected to gas chromatography and that several types of derivatives are useful for this purpose. Isopropylidene derivatives should be useful in the gas chromatography of various antimetabolites and synthetic intermediates, but acetyl, trimethylsilyl, and probably benzyl derivatives can be utilized more generally for all nonphosphorylated forms of the vitamin. We have been completely successful in resolving the problems of quantitation with respect to the acetyl derivatives. The formation of trimethylsilyl derivatives, however, created problems; but these derivatives are especially attractive, since such derivatives of several phosphorylated forms of vitamin B_6 have been shown to be amenable to gas chromatography, and thus they offer promise for the development of a rapid and comprehensive assay for all forms of vitamin B_6.

The sensitivity of the methods depends mainly on the instrumentation. With our instrument, we have been able to determine 1 μg or less of each vitamer. In this connection, trifluoroacetyl derivatives may offer some advantages, such as the possibility of utilizing the more sensitive electron capture detectors, as well as greater volatility; but such derivatives have not yet been studied.

Likewise, the applicability of gas chromatography to vitamin B_6 compounds in biological material remains to be determined. Mass spectrometry of these compounds is under active study, particularly because it would offer, with appropriate instrumentation [6], a means for qualitative determination of effluents from the gas chromatography column. All vitamin B_6 compounds studied so far, with the exception of phosphorylated and N-methylated derivatives, have given molecular ions [22]. Although the mass spectra of only the isopropylidene derivatives have been published [22], tri-O-acetylpyridoxol has also been found to give a characteristic spectrum. Finally, detailed consideration of the gas chromatography of the vitamin B_6 compounds should help to indicate the capabilities and limitations of the method for other vitamins, a research area that can stand a good deal of study.

OTHER WATER-SOLUBLE VITAMINS

Nicotinamide has a structure related to that of vitamin B_6. Its considerably greater simplicity should make it amenable to gas chromatography involving a number of different columns and conditions. Nevertheless, I am not aware of any application of gas chromatography for assaying nicotinamide in biological material, although conditions for the gas chromatography of ethyl ester of nicotinic acid have been described [23].

L-Ascorbic acid (vitamin C) forms a TMS derivative under standard conditions [19] and gives a single peak with a thin-film (SE-32) [23] or thick-film (diethylene glycol succinate) [24] column.

Myoinositol, whose significance as a vitamin is still debated, likewise has been chromatographed as a TMS derivative on a thin-film (SE-32) column (along with other isomeric inositols) [19] and on a thick-film (diethylene glycol succinate) [24] column. A gas chromatographic procedure for estimating myoinositol in yeast has been worked out [24].

Biotin and pantothenic acid have structures sufficiently simple to invite study by gas chromatography. Presumably, these vitamins could be subjected to gas chromatography as suitable derivatives (e.g., esters or TMS derivatives).

Folic acid, thiamine (vitamin B_1), riboflavin, and vitamin B_{12} are less inviting for the gas chromatographer. Here the possibility exists for the application of pyrolytic procedures in conjunction with gas chromatography of the resulting fragments. Alternatively, some of these vitamins, especially folic acid and thiamine, could be cleaved by chemical or enzymatic procedures to yield simpler but well-defined fragments that would be amenable to gas chromatography.

FAT-SOLUBLE VITAMINS

Gas chromatography has been tried for all of the fat-soluble vitamins, and most of them have been found amenable to the technique.

Gas chromatography of all-trans-retinol (vitamin A) and its acetyl derivative has been studied with SE-20 and QF-1-0065 columns (3% on Gas Chrom P) at 165 or 181° column temperature [25]. Both the parent compound and its acetyl derivative gave multiple peaks. It has been suggested that alterations in the conjugated double bonds occur during gas chromatography; hydrogenation of the vitamin gave a single main peak of lowered retention time [25].

Vitamin D was one of the first vitamins to be studied by gas chromatography. Ziffer et al. [26] found that both vitamins D_2 and D_3 are thermally transformed into tetracyclic systems as "pyro" isomers, which can be separated on an SE-30 thin-film column. The relative amounts of the "pyro" isomers of both vitamins are reproducible and are only slightly dependent on the "flash heating" temperature. This observation has been confirmed by Nair et al. [27], who applied the method to the quantitative estimation of vitamins D_2 and D_3 in biological material. The latter investigators also studied the gas chromatography of vitamins D_2 and D_3 as their trifluoroacetyl and TMS derivatives, utilizing various columns. Gas chromatography of the pro-vitamins D apparently gave no problems on thin-film columns (SE-30 and neopentyl glycol succinate) [26].

Gas chromatography of vitamin E (tocopherol) has been studied in several laboratories [25, 29, 30], using thin-film columns and various derivatives. The tocopherol series comprises at least eight naturally occurring compounds, and the best chemical methods for analysis of mixtures of them are rather involved. Wilson et al. [25] carried out a thorough study of the gas chromatography of the parent compounds and their acetyl derivatives on several columns [SE-30, QF-1, and poly-(ethylene glycol adipate)]. The retention times increased regularly with the number of additional methyl groups on the tocol moiety and showed a straight-line relationship between the logarithm of the retention time and the number of carbon atoms, a relationship similar to that found for homologous pyridoxol derivatives (Fig. 1). Some positional isomers could not be resolved. As little as 0.2 μg of tocopherol could be estimated. The two tocotrienols, having three double bonds in the side chain, had longer retention times than the fully saturated tocol derivatives. The method was applied to the assay of tocopherols; the results were found to be in agreement with an independent method of assay. Nair and Turner [30] studied the gas chromatography of tocopherols and their TMS derivatives on silicone polymer columns. They could separate α-tocopherylquinone and α-tocopherylhydroquinone from the original α-tocopherol. α-Tocopherol has also been assayed in animal tissue by gas chromatography [31].

Carrol and Herting [25] have applied gas chromatography to vitamin K_1. The compound had a considerably longer retention time than any of the tocopherols. Two peaks were obtained on an SE-30 column, with retention times of 41.7 and 50.3 min, respectively, and peak areas in a ratio of 4:1. On a QF-1-065 column, the same preparation gave a single peak with a retention time of 27.8 min. It is not known whether the second peak observed on the SE-30 column was due to an impurity in the preparation or whether it resulted from an alteration in the compound on the column. The vitamin K_2 molecule, with its 41 carbon atoms, is probably too large for gas chromatography under these conditions [25]. The same applies to ubiquinones (coenzyme Q), the retention times of which have been estimated to exceed 12 hr [25].

Finally, the essential fatty acids have also been regarded as a vitamin ("vitamin F"). Gas chromatography has been long established as the analytical method of choice for these compounds, and this use of the procedure has been adequately reviewed elsewhere [32].

ADDED IN PROOF

Continued work on the gas chromatography of vitamin D has resulted in improvements. Murray et al. [33] have developed an assay method in which vitamins D_2 and D_3 are converted into "isovitamins D" by treatment with antimony trichloride and tartaric acid. The "isovitamins" give single and separate peaks on columns packed with Celite coated with 3% silicone oil. The method was combined with initial purification by thin-layer chromatography and gave results in reasonable agreement with other assay procedures. Avioli and Lee [34] estimated vitamins D_2 and D_3 as their "pyro" derivatives [27] on an SE-30 thin-film column and have applied the method to purified plasma samples. They detect 5 ng, which represents a substantial increase in sensitivity over spectrofluorimetric techniques, as well as over the gas chromatographic methods previously reported [33].

ACKNOWLEDGMENT

The principal investigation cited in this review was supported in part by U.S.P.H.S. Research Grants Nos. CA-05697 and CA-08793 from the National Cancer Institute.

REFERENCES

1. R. Strohecker and H. M. Henning, Vitamin Assay, Verlag Chemie, 1965.
2. E. Heftmann, Chromatography, Reinhold, New York, 1960, pp. 645, 473, and 597.
3. C. A. Storvick and J. M. Peters, Vitamins Hormones, 22:833, 1964.
4. C. A. Storvick, E. M. Benson, M. A. Edwards, and M. J. Woodring, Methods Biochem. Anal. 12:183, 1964.
5. E. W. Toepfer and M. M. Polansky, Vitamins Hormones 22:825, 1964.
6. S. W. Downer, in: L. R. Mattick and H. A. Szymanski., eds., Lectures on Gas Chromatography, Plenum Press, New York, 1965, p. 189.
7. E. E. Snell, Vitamins Hormones 16:77, 1958.
8. S. Shaltiel, E. H. Fischer, and J. L. Hedrick, Biochemistry 5:2109 1966.
9. B. E. Haskell and E. E. Snell, Arch. Biochem. Biophys. 112:494, 1965.
10. E. E. Snell et al., Chemical and Biological Aspects of Pyridoxal Catalysis, Macmillan, New York, 1963.
11. W. Korytnyk, J. Med. Chem. 8:112, 1965, and W. Korytnyk and B. Paul, J. Heterocyclic Chem. 2:144, 1965. References to earlier papers in the series are given in these two papers.

12. W. Korytnyk, G. Fricke, and B. Paul, Anal. Biochem., 17:66, 1966.
13. W. Korytnyk and W. Wiedeman, J. Chem. Soc., p. 2531, 1962.
14. R. Kuhn and G. Wendt, Ber. 71:780, 1938.
15. W. Korytnyk, J. Org. Chem. 27:3721, 1962.
16. W. Korytnyk and B. Paul, Tetrahedron Letters, No. 8, p. 777, 1966.
17. After the principal study cited in the present paper was completed, a preliminary report on the gas chromatography of some acetyl derivatives of vitamin B-compounds appeared: A.R. Prosser and A.J. Sheppard, Federation Proc. 25:669, 1966.
18. T. Hashizume and Y. Sasaki, Anal. Biochem. 15:199, 1966.
19. C.C. Sweeley, R. Bentley, M. Makita, and W.W. Wells, J.Am. Chem. Soc. 85:2497, 1963.
20. S. Friedman and M.L. Kaufman, Anal. Chem. 38:144, 1966.
21. B. Paul and W. Korytnyk, Abstracts of the 152nd Meeting of the American Chemical Society, September, 1966, p. 63.
22. D.C. DeJongh, S.C. Perricone, and W. Korytnyk, J. Am. Chem. Soc. 88:1233, 1966, and unpublished results.
23. H. Kubiezkova, V. Rezl, and N. Kucharczyk, Abstracts, J. Gas Chromat. 1, No. 4, 1963.
24. R.N. Roberts, J.A. Johnston, and W.B. Fuhr, Anal. Biochem. 10:282, 1965.
25. K.K. Carroll and D.C. Herting, J. Am. Oil Chem. Soc. 41:473, 1964.
26. H. Ziffer, W.J.A. Van den Heuvel, E.O.A. Haahti, and E.C. Horning, J. Am. Chem. Soc. 82:6411, 1960.
27. P.P. Nair, C. Bucana, S. de Leon, and D.A. Turner, Anal. Chem. 37:631, 1965.
28. N. Nicolaides, J. Chromat. 4:496, 1960.
29. P.W. Wilson, E. Kodicek, and V.H. Booth, Biochem. J. 84:524, 1962.
30. P.P. Nair and D.A. Turner, J. Am. Oil Chem. Soc. 40:353, 1963.
31. J.G. Bieri and E.L. Andrews, Iowa State J. Sci. 38:3, 1963.
32. W.R. Supina, in: H.A. Szymanski, ed., Biomedical Applications of Gas Chromatography, Plenum Press, New York, 1964, p. 271.
33. T.K. Murray, K.C. Day, and E. Kodicek, Biochem. J. 98:293, 1966, 96:29P, 1965.
34. L.V. Avioli and S.W. Lee, Anal. Biochem. 16:163, 1966.

Dual Channel Gas Chromatography

Ernest J. Bonelli

Varian Aerograph
Walnut Creek, California

The term "dual channel gas chromatography" refers to a special class of multichannel gas chromatography. The term as we use it today means the use of two detectors for examining the column effluent. It was first used by Dr. James Lovelock, who has shown simultaneous chromatograms of chlorinated substances obtained with an electron capture detector and an argon detector [1]. In addition to showing two chromatograms, dual channel gas chromatography offers more than twice the qualitative and quantitative information than is available in a single channel system.

Some of the available detector combinations follow:

1. Thermal conductivity (TC) and flame ionization (FI) [2]. The TC detector is sensitive to all compounds, except the carrier gas; however, its sensitivity is limited. The FI detector is 1000 times more sensitive than the TC; however, it shows little or no sensitivity to fixed gases. This combination is employed in the analysis of ppm hydrocarbons in fixed gases.

2. Thermal conductivity and infrared, ultraviolet, NMR, or mass spectrometry. Gas chromatography is used to separate and quantitate; the other detectors are used to identify the pure compounds.

3. Thermal conductivity and radioactive counting. Here the the radioactive detector is used to determine which of the eluting peaks are radioactive. This detector combination is generally connected in series.

4. Thermal conductivity and color tests for functional groups [3]. Here one bubbles the effluent into reagents, which gives a color reaction or precipitate, depending on the functional groups present.

5. Multiple electron capture (EC) detectors [4]. Operated at various voltages to give selective detection.

6. Flame ionization and electron capture detectors. Here the FI detector is used as a mass detector, whereas the EC detector response is related to a specific property, the electron affinity, of the compound being analyzed. The electron affinity, the extent to which a compound will adsorb electrons, may vary greatly—from 1 for saturated hydrocarbons to about 400 million for a compound like carbon tetrachloride.

7. Flame ionization and phosphorus detector. Here also the FI detector is the mass detector and the phosphorus detector responds to the specific element, phosphorus.

8. Phosphorus and electron capture detectors. This interesting combination has been employed in pesticide analysis.

In order to operate such multichannel systems, there are many detector characteristics and operating parameters to be considered. Some of these follow:

1. Sensitivity. Because of sensitivity differences, it is entirely possible that one detector will not "see" the sample whereas a second detector will give a large response. Figure 1 shows a dual channel trace of the pesticide aldrin, 95% pure. The sample, as it elutes from the chromatographic column, is split 50-50. The two streams are sent to an electron capture and flame ionization detector, respectively. The flame detector, a mass detector, indeed shows the sample to be 95% pure; however, it does not detect the trace impurities shown by the ultrasensitive electron capture detector.

2. Linear range. If area ratios are to be measured, it is important that the detectors are operated within their linear range, that is, the detector response should be a linear function of the sample injected. For example, in the EC detector the linear range is quite narrow, 10^3. The FI detector, however, has a linear range of 10^6.

3. Carrier gas. Carrier gas should be another consideration. Some detectors work well only with specific gases. Thermal conductivity detectors operate best using helium as carrier gas. The argon detector, of course, requires only argon. For the electron capture—flame ionization combination, we have found nitrogen to be the best carrier gas.

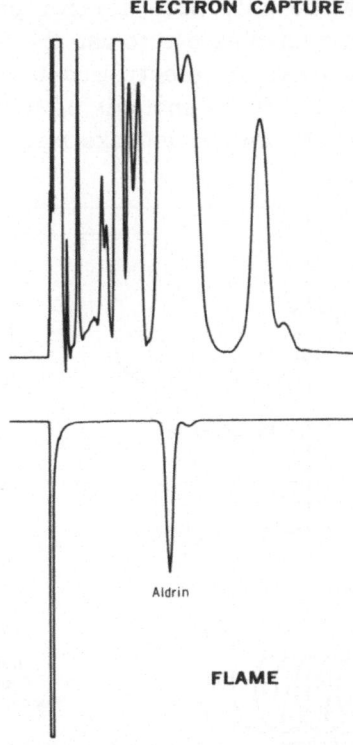

Fig. 1. Dual channel chromatogram of aldrin—95% pure.

EC/FLAME COMBINATION

A combination we have studied at great length is the use of the flame ionization and electron capture detectors. This is shown in Fig. 2. The effluent of the column is sent into two detectors by means of a tee leading to two restrictors. The restrictors are used to aid the matching of the detectors and also as a means of varying the effluent split. These are usually short sections of capillary tubing. One of the detectors could be switched completely off pneumatically by introducing a carrier gas stream at the entrance to the restrictor. Thus, in using an electron capture–flame ionization combination, carrier gas may be introduced on the electron capture side of the split at such a pressure that all of the sample goes through the flame detector while only pure carrier gas goes through the EC detector. This enables one to use the flame unit alone and at the same time to keep the electron capture cell purged. This is particularly useful in the analysis of aqueous samples. The electron capture detector cannot take large injections of water,

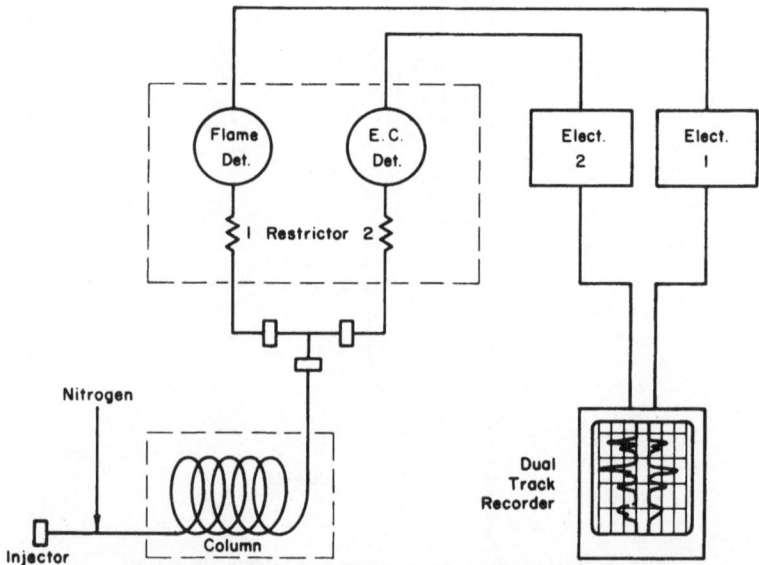

Fig. 2. Schematic diagram of a dual channel chromatograph and recorder.

Electron Capture

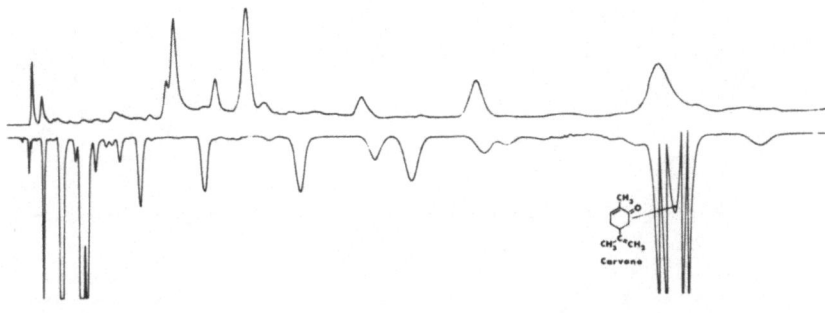

Flame Ionization

Fig. 3. Dual channel chromatogram of spearmint oil.

whereas water is insensitive to the flame ionization detector.
The electron capture detector can then be switched off pneumatically while the water peak goes through the flame ionization
detector.   After the water has eluted, the electron capture
detector can be used to identify the remaining compounds in
this aqueous solution.   This technique may be used to adjust
split ratios, although in practice it is usually done by changing
restrictors.

The data obtained by the two detectors can be presented
on two recorders, one for each detector. However, working
with a complex mixture, it is necessary to line up the recorder
traces exactly, and this is not an easy job. The method of
choice would be to use a two-pen recorder with a different
color ink for each pen.

Perhaps the most-used application of dual channel gas
chromatography is in the analysis of foods, flavors, and natural
products. As one can imagine, these are extremely difficult
to analyze, and it seems logical to apply this method to attempt
to identify some of the constituents.

Spearmint and peppermint oils were the first compounds
upon which we employed dual channel gas chromatography [5].
Figure 3 is a dual channel trace of spearmint oil. As you can

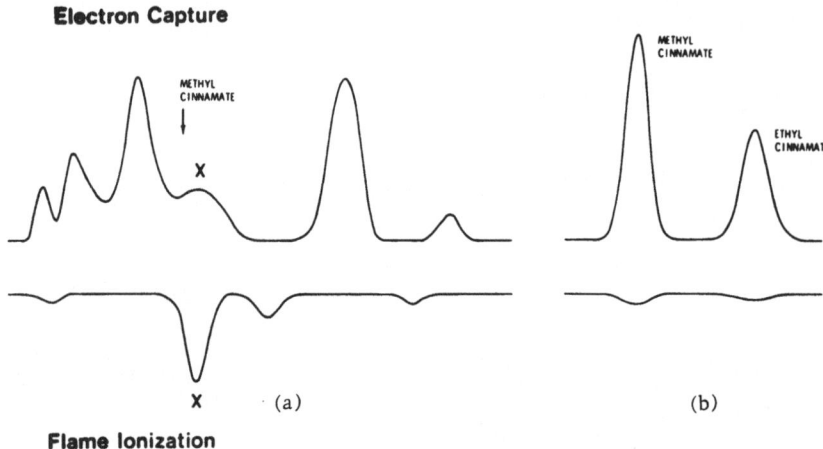

Fig. 4. (a) High boiling components of peppermint oil. (b) Dual channel response to methyl and ethyl cinnamate.

see, there are peaks on the EC side that have no counterpart on the flame ionization trace and vice versa.

Figure 4 shows a trace of high-boiling components of peppermint oil and indicates a peak X on both flame and EC. This is the only component that shows a simultaneous peak from both detectors. The area is approximately equal for both the flame and the EC. At first, this peak was thought to be methyl cinnamate and the basis of retention time. Further research, however, showed that this could not possibly be so because the sensitivity (area) of the methyl cinnamate to EC is approximately 100 times greater than to flame. This ratio of peak areas then can be an extremely valuable parameter for the verification of an unknown compound.

The ratio of electron capture response to flame response (EC/flame) is designated as ϕ [6]. This is an empirical value based on $1X$ attenuation for both electrometers and assumes a 1:1 split in the column effluent. The ϕ ratio is not an absolute parameter. Its value for a particular compound cannot be expected to be constant for all chromatographic conditions. Contributing most to this inconsistency is the response of the electron capture detector.   This response is dependent upon the standing current that in turn can be affected by column bleed and foil contamination.

In practice, variability of ϕ is not a serious problem. From day to day, the ϕ values may change by about 10%, but this variation can be compensated by normalizing to a standard. Within one set of chromatographic conditions the reproducibility of ϕ values is as good as can be expected from any chromatographic analysis.   Table I shows an example of typical reproducibility for three compounds, di-n-propyl-, allyl propyl-, and di-allyl disulfides on five successive injections.

Another project undertaken was to analyze electron capture components in cruciferae [7]. The vegetables chosen were watercress, horse radish, rutabaga, radish, turnip, and white cabbage. An extract of rutabaga is shown in Fig. 5. One of the components, peak No. 39, was confirmed by infrared to be 2-phenylethyl isothiocyanate. The column used in this analysis was a 10 ft $\times$ $\frac{1}{8}$ in. stainless steel column packed with 10% Carbowax 20 M on 60/80 mesh Chromosorb W. Other conditions were: column temperature, 150°C; nitrogen flow rate, 50 ml per min (split ratio 1:1); hydrogen flow rate, 25 ml per min. The instrument used was Varian Aerograph Model No. 204-1B with a two-channel Westronics 1 mV recorder.

It was thought that the sense of smell might somehow be related to electron affinity. We chose to analyze a clove of garlic. The clove was placed in a small jar, and a 1 ml headspace sample was injected in a two-channel gas chromatograph. The resulting trace is shown in Fig. 6. As you can see, the

Table I. Reproducibility of ϕ^a

Di-n-propyldisulfide	Allyl propyl disulfide	Diallyl disulfide
1.84	13.4	22.8
1.81	12.8	23.3
1.76	13.0	23.8
1.58	12.2	22.9
1.78	13.0	22.9

$^a\phi$ for three compounds at 145°C with 10 ft, 10% Carbowax 20 M column.

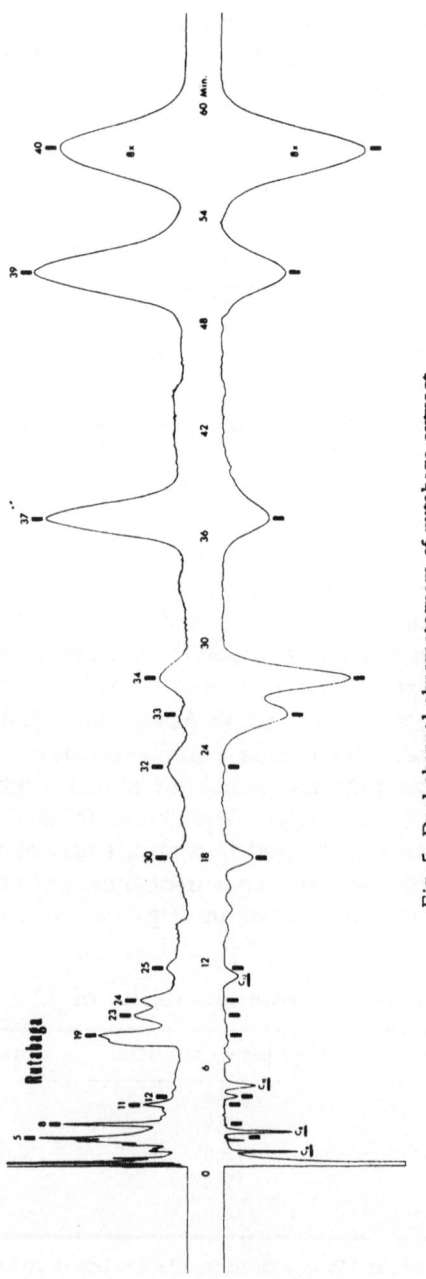

Fig. 5. Dual channel chromatogram of rutabaga extract.

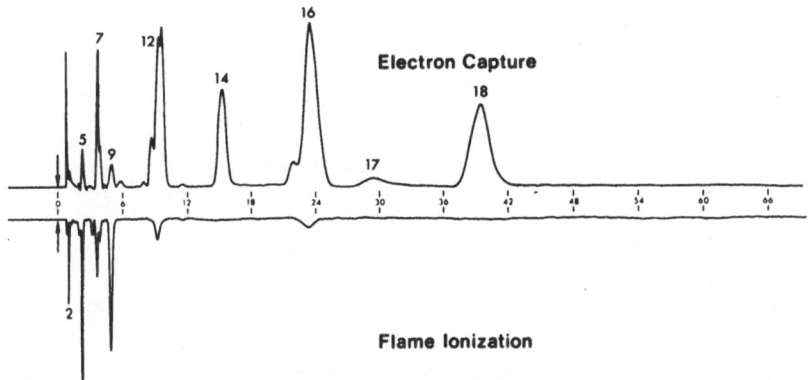

Fig. 6. Dual channel response to garlic headspace.

chromatogram shows many very large electron capture peaks with virtually no flame response. The peaks were identified as: No. 2, methyl mercaptan; No. 5, methyl allylsulfide; No. 7, dimethyl disulfide; No. 9, diallylsulfide; No. 12, methyl allyl-disulfide; No. 14, dimethyl trisulfide; No. 16, diallyldisulfide; No. 17, methyl n-propyl trisulfide; No. 18, methyl allyl-trisulfide.

An extract of onion was also analyzed and was found to contain only about 10% of the amount of sulfide compounds in garlic. This is some analytical truth as to why garlic is so much more potent than onion. The sulfides contained in onion are mostly symmetrical and mixed methyl and propyl sulfides.

Some of the other compounds we have analyzed are shown in the following figures. Figure 7 shows imitation oil of nutmeg. A 5 ft × ⅛ in. stainless steel column containing 30% FFAP on 70/80 DMCS-treated Chromosorb W was used in this analysis. The column temperature was 145°C; the injector temperature, 220°C; the detector temperature, 200°C.

The next chromatogram (Fig. 8) is of cognac oil. This was analyzed with a 10 ft × ⅛ in. stainless steel column with 12.5% Carbowax 20 M on 60/80 HMDS-treated Chromosorb W. The column temperature was 176°C; the injector, 200°C; the detector, 205°C.

Figure 9 shows a chromatogram of lemon oil. The column used in this analysis was 12.5% Carbowax 20 M on

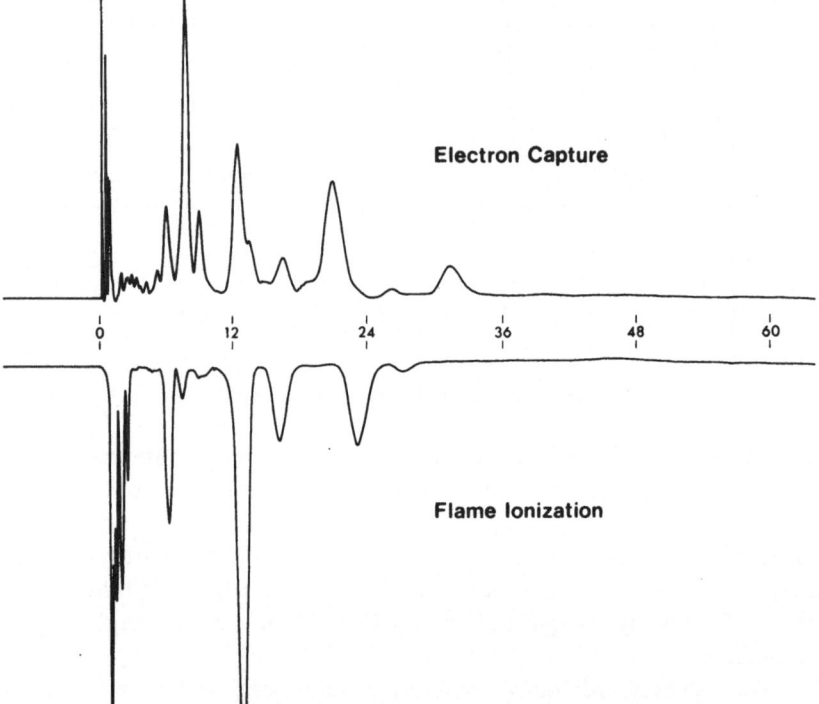

Fig. 7. Dual channel chromatogram of nutmeg.

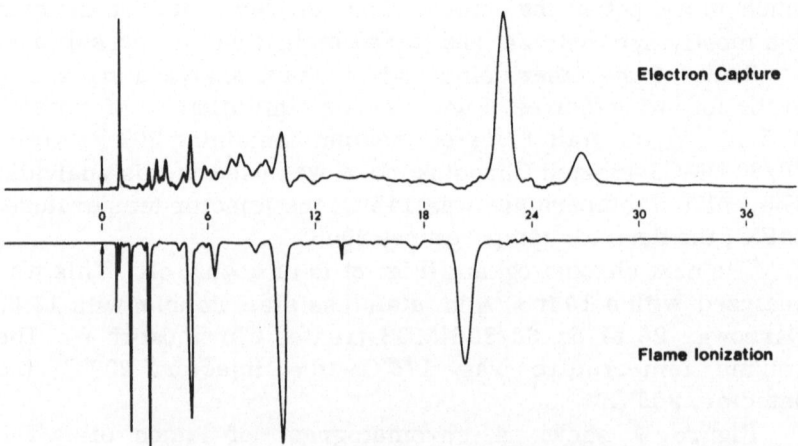

Fig. 8. Dual channel chromatogram of cognac.

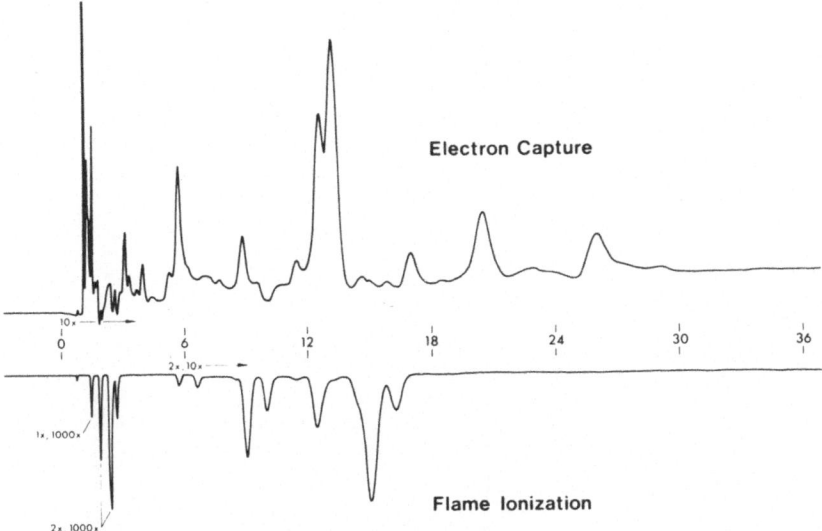

Fig. 9. Dual channel chromatogram of lemon oil.

60/80 HMDS-treated Chromosorb W. The column temperature was 150°C; the injector, 200°C; the detector, 175°C.

Still another dual channel trace is shown in Fig. 10. This is a coffee extract analyzed with a 10 ft × $\frac{1}{8}$ in. 12.5% Carbowax 20 M HMDS-treated Chromosorb W column. The column temperature was 139°C; the injector, 180°C; the detector, 145°C.

Dual channel gas chromatography can also be used to determine differences between natural and synthetic peppermint oil. Figures 11 and 12 show synthetic and then natural peppermint oil. Notice the preponderance of compounds in the natural oil.

EC/PHOSPHORUS COMBINATION

Another recent detector combination [8] shown in Fig. 13 is the EC/phosphorus. This has been employed in pesticide analysis in the case of string bean forage (Fig. 14). The electron capture trace shows many peaks which, without

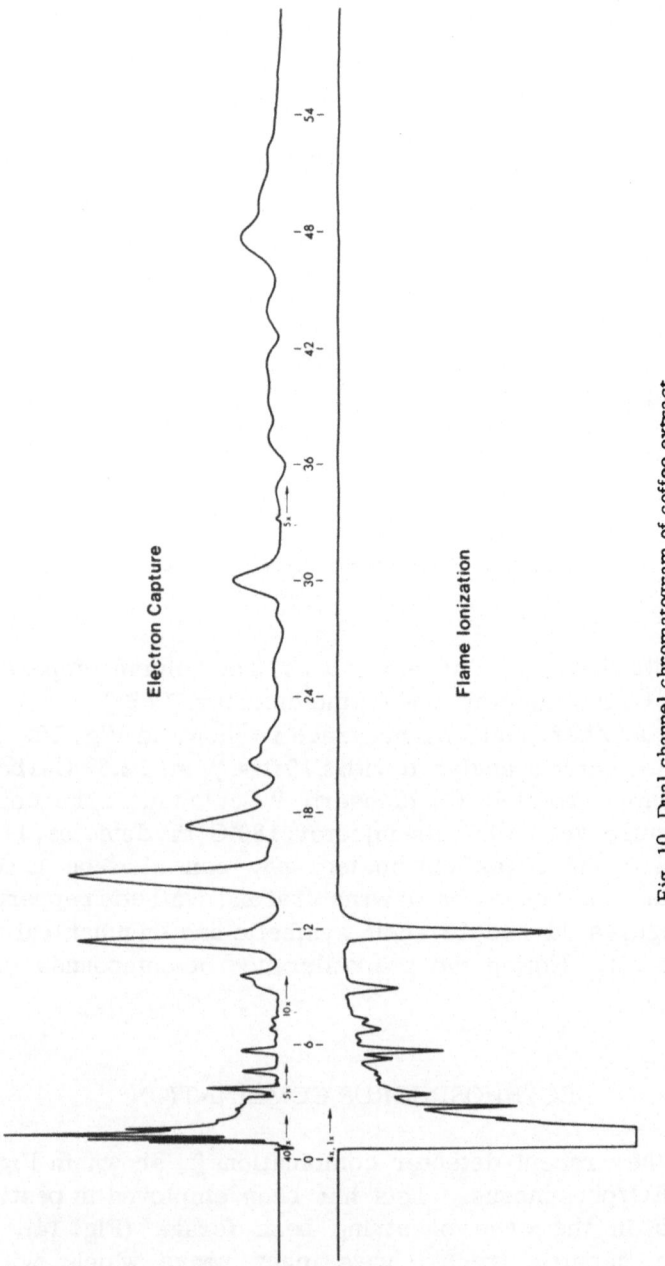

Fig. 10. Dual channel chromatogram of coffee extract.

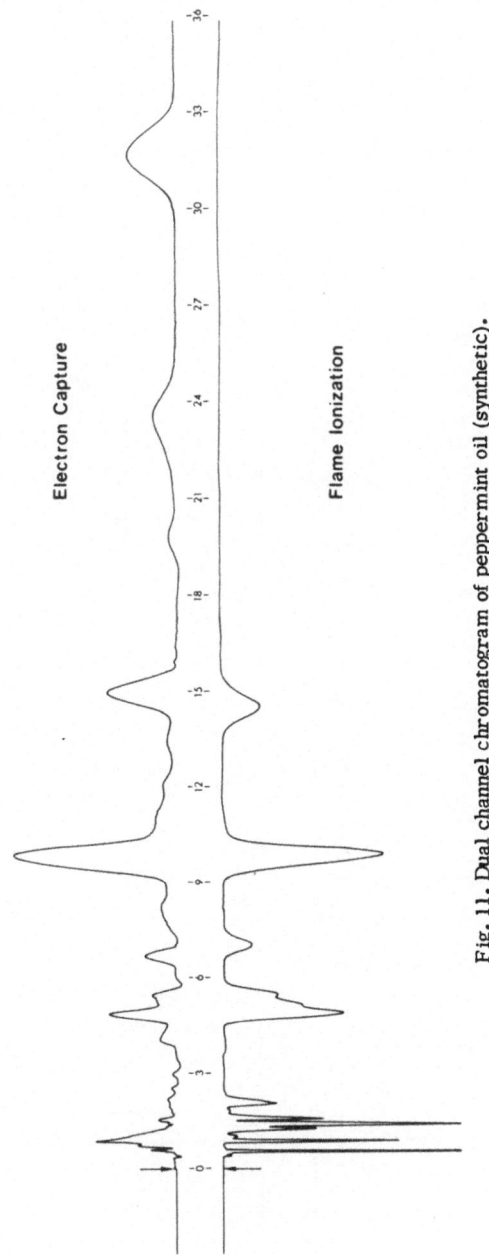

Fig. 11. Dual channel chromatogram of peppermint oil (synthetic).

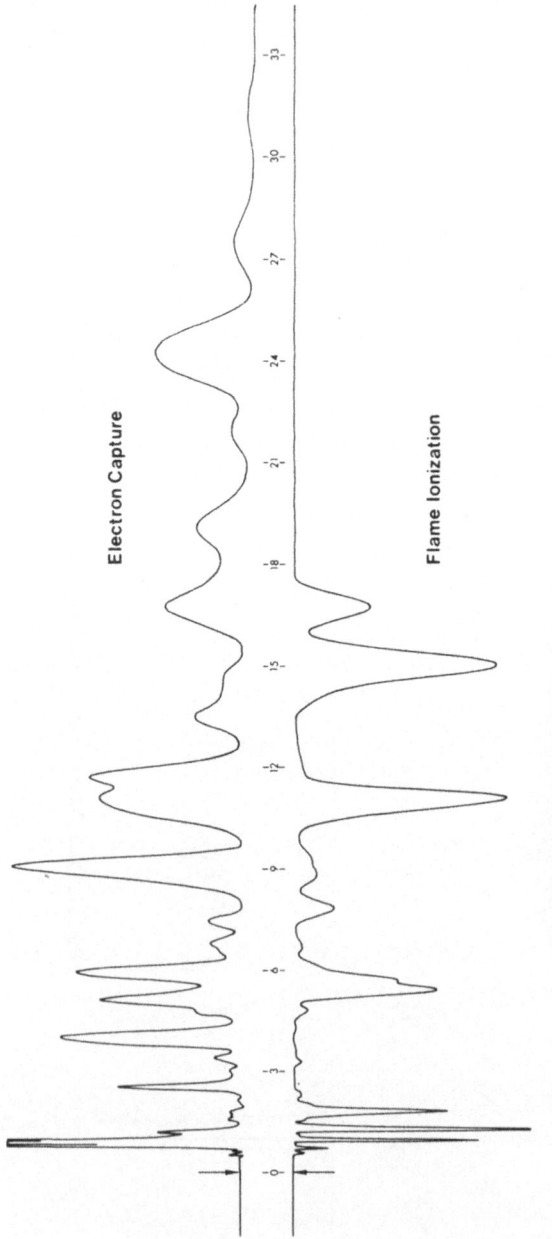

Fig. 12. Dual channel chromatogram of peppermint oil (natural).

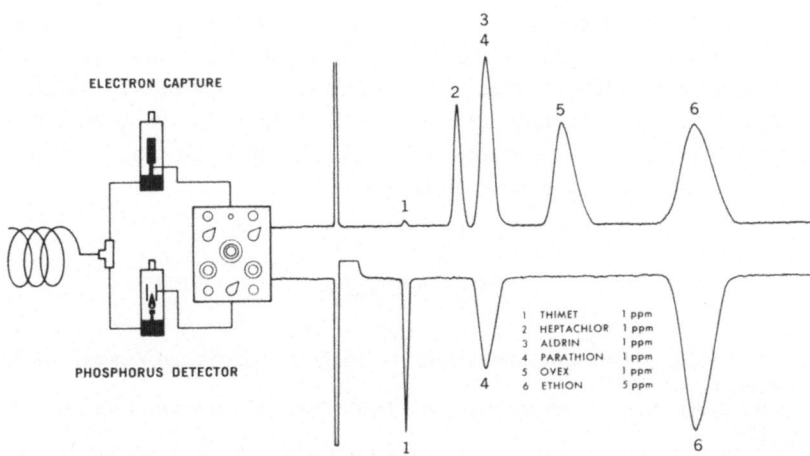

Fig. 13. Schematic and dual channel EC/P chromatogram of a standard pesticide mixture.

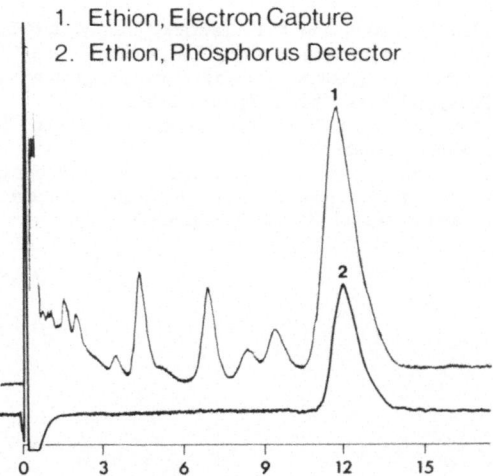

Fig. 14. Dual channel chromatogram of ethion in string bean forage.

additional information, would surely confuse the residue analyst. The phosphorus detector, however, disregards all the compounds except the organo phosphate pesticide ethion.

Dual channel gas chromatography offers a new parameter for the identification of unknown compounds. It may complicate an already complicated field, but it gives the analyst another dimension with which to work and another device to add to his research budget.

REFERENCES

1. J.E. Lovelock, "Ionization Methods for the Analysis of Gases and Vapors," Anal. Chem. 33:162, 1961.
2. D.W. Grant, Gas Chromatography, ed. D.M. Desty, Butterworth, London, 1958, p. 153.
3. J.T. Walsh and C. Merrit, Jr., "Qualitative Functional Group Analysis of Gas Chromatographic Effluents," Anal. Chem. 32:1378, 1960.
4. J.E. Lovelock and S.R. Lipsky, "Electron Affinity Spectroscopy—A New Method for the Identification of Functional Groups in Chemical Compounds Separated by Gas Chromatography," J. Am. Chem. Soc. 82:431, 1960.
5. C.H. Hartmann, D. Oaks, and K.P. Dimick, "Essential Oil Analysis by Two-Channel Gas Chromatography," American Chemical Society Meeting, New York, September, 1963.
6. D.M. Oaks, C.H. Hartmann, and K.P. Dimick, "Analysis of Sulfur Compounds with Election Capture/Hydrogen Flame Dual Channel Gas Chromatography," Second International Symposium on Gas Chromatography, Houston, Texas, March 23-26, 1964, and Anal. Chem. 36:1560, 1964.
7. F. Tao, "E.C. Components in Cruciferae" (Summer 1964) and "Isothiocyanates" (Fall 1964), Aerograph Research Notes.
8. D.M. Oaks, K.P. Dimick, and C.H. Hartmann, Varian Aerograph Seminar, Los Angeles, California, May 6, 1966, and C.H. Hartmann, "Aerograph Phosphorous Detector" Aerograph Research Notes, Spring, 1966.

Analytical Methods for Pesticides

H. P. Burchfield

Pesticide Research Laboratory
U. S. Public Health Service
Perrine; Florida

INTRODUCTION

The analysis of pesticide residues presents unusual problems because of the small amounts of materials that must be measured and the large amounts of interfering substances that must be removed. Tolerances for all compounds used on edible crops are set by the Food and Drug Administration. For compounds that possess low toxicity, tolerances are relatively high. Thus, residues of 7 ppm of DDT are permissible. However, crops containing a few hundredths of a ppm of endrin can be condemned. It is often necessary to analyze for much smaller residues than these in environmental health studies, since residues of pesticides in water at the parts per trillion (ppt) level are significant because aquatic flora and fauna are able to accumulate them against concentration gradients. In general, it is necessary to be able to analyze for pesticides at the 0.01 to 10 ppm level in foods and animal tissues and at the ppt or ppb levels and upward in environmental pollution studies.

SAMPLE PREPARATION

Within recent years very sensitive methods of detection have been developed. Even so, it is necessary to concentrate samples to a high degree prior to analysis. At the same time, the pesticides must be isolated from various interfering materials that occur in the environment. The procedures by which these objectives are achieved are called extraction and cleanup. These operations are basic to residue analysis, and must be accomplished quantitatively for the results to be meaningful. The methods described below are particularly applicable to chlorinated hydrocarbons and allied compounds. Modified procedures must often be employed for the analysis of organic phosphates, pesticide metabolites, and herbicides of the 2, 4-D class. (See "Guide to the Analysis of Pesticide Residues" [1] for details of methods.)

Extraction

The first step in preparing a sample for analysis is to extract the pesticide from it with an organic solvent. The solvent used may be polar, nonpolar, or a mixture of polar and nonpolar solvents. The choice will depend upon the nature of the sample and the polarities of the pesticides to be extracted. After extraction, the organic solvent containing the pesticides is evaporated to recover the residue. The residues obtained from water (1 liter) and tissue (100 g) samples may weigh as little as 1 mg to 1 g. This reduction in mass is made possible by the fact that many samples (excluding soils, fats, etc.) contain 90% or more water, most of which does not partition into the organic solvent. Also, many cellular constituents including cellulose, proteins, amino acids, carbohydrates, and related compounds are sparingly soluble in most organic solvents. However, most pesticides dissolve in polar and/or nonpolar organic solvents and hence are separated from the bulk of the sample. On evaporation of the organic solvent the pesticides are concentrated. This reduces the amount of material that must be handled in subsequent steps of the analytical procedure.

Cleanup

Many compounds other than pesticides that can be present in water, soil, and tissues are extractable with organic solvents.

These include industrial pollutants, lipids, pigments, and other nonpolar to moderately polar compounds. It is sometimes possible to analyze for pesticides in water and soil extracts without further separation from other compounds. With most extracts, cleanup (purification) is required to reduce the weight of the residue and remove substances that interfere in the analytical step. A variety of cleanup procedures have been utilized, but only a few are universal enough in scope to be applied to a wide variety of situations. These are cleanup by partition, by chromatography, and by saponification. In the partition method, the residue obtained after extraction is shaken in a separatory funnel with a mixture of two immiscible solvents. A nonpolar solvent (usually hexane or petroleum ether) and a polar solvent (usually acetonitrile) are always employed. Fatty materials dissolve in the nonpolar solvent, while most pesticides are partitioned into the polar phase. The two layers are then separated, and the pesticide residue is recovered by evaporation of the polar solvent. This procedure is very useful for the separation of pesticides from large amounts of oils and fats. However, it is not efficient for the removal of polar lipids, plant pigments, and other interfering compounds. Consequently, an adsorption step is usually required to remove these materials. This is carried out by dissolving the residue in an organic solvent and adding adsorbent or by percolating the solution through a column of adsorbent packed in a chromatographic column. Generally, adsorbents are chosen that possess high affinities for plant pigments and related compounds, but do not adsorb pesticides.

Samples that contain large amounts of fats and oils can be cleaned up efficiently by saponification. The samples are digested with alcoholic potassium hydroxide to convert triglycerides to soaps and glycerine. The solution is quenched with water and the pesticides extracted with an organic solvent such as petroleum ether. The soaps remain in the aqueous phase, which is discarded. This method is not as generally applicable as partition and column chromatography since many pesticides are destroyed by alkali. However, samples containing dieldrin, endrin, and aldrin can be cleaned up by this method. DDT is dehydrochlorinated to DDE and can be measured as such.

DETERMINATIVE METHODS

After the pesticide has been extracted from the sample and interfering compounds removed, the residue is ready for analysis. Formerly, specific methods were used for each individual pesticide in order to establish its identity and determine the amount present. These are usually colorimetric tests that are based on causing the pesticide to react with a reagent to produce a chromophore. The intensity of the color is then measured and the amount of pesticide present is computed from the color obtained on solutions of standards of known concentration.

During the past ten years colorimetric methods have been gradually superseded by chromatographic procedures in which a number of pesticides can be analyzed in the same operation. Chromatography per se is a technique for separating complex mixtures of compounds rather than analyzing for them quantitatively. However, when chromatography is used in conjunction with quantitative measuring systems, the combined operation provides more information to the residue analyst than any other technique available. Three forms of chromatography are currently in use: paper, thin-layer, and gas. The use of paper chromatography is declining while that of thin-layer chromatography is increasing. However, gas chromatography is superior to both of these methods with respect to resolution and sensitivity.

Paper Chromatography

To carry out paper chromatography, the sample is extracted and purified as described above and an aliquot of the residue is spotted near one end of a strip of filter paper. The end (lower) of the filter paper nearest the point where the sample was spotted is then dipped in a reservoir containing an organic solvent, the reservoir and filter paper being contained in an airtight jar or box. The solvent moves up the filter paper by capillarity, carrying the pesticide with it. In most cases, the pesticides move up the paper at slower rates than the solvent because they are partitioned between the stationary phase (paper) and the mobile phase (solvent). Also, the individual pesticides are usually carried along by the solvent at different rates because their partition coefficients between the stationary phase and mobile phase vary from compound to compound. The paper is removed from the jar or chromatographic cabinet shortly before the solvent front reaches the top of the paper. By this time, the pesticides are spread out along the longitudinal axis of the paper as discrete spots. The distance between the point of sample application and the center of each spot, divided by the distance between the point of application and the solvent front at the termination of the run, gives the R_f value of the compound. Each pesticide possesses a characteristic R_f value under any given set of chromatographic conditions. However, some of these may approach each other closely. If the R_f values are widely separated, resolution will be good; if not, it will be poor. The resolution achieved for any pair of pesticides can be adjusted by changing the solvent system or pretreating the paper. An example of the separation of pesticides by paper chromatography is shown in Fig. 1.

It is not possible to detect the presence of compounds on the paper unless it is chemically treated. Usually, the paper is sprayed with a reagent that is specific for groups of compounds. Thus, a spray containing silver ion will serve to detect chlorinated hydrocarbons. The silver ion reacts with these compounds to yield silver chloride, which darkens on exposure to light. Similarly, specific reagents can be used for the detection of organic phosphate insecticides. Thus, it is possible to separate a series of related compounds on paper and to detect them all by spraying the paper with a reagent. This makes it possible to analyze simultaneously for a number of compounds even when the history of the sample is unknown. Paper chromatography is extremely useful as a screening test and has some advantages over gas chromatography for this purpose. The equipment is inexpensive, and a large number of samples can be run simultaneously in a single cabinet.

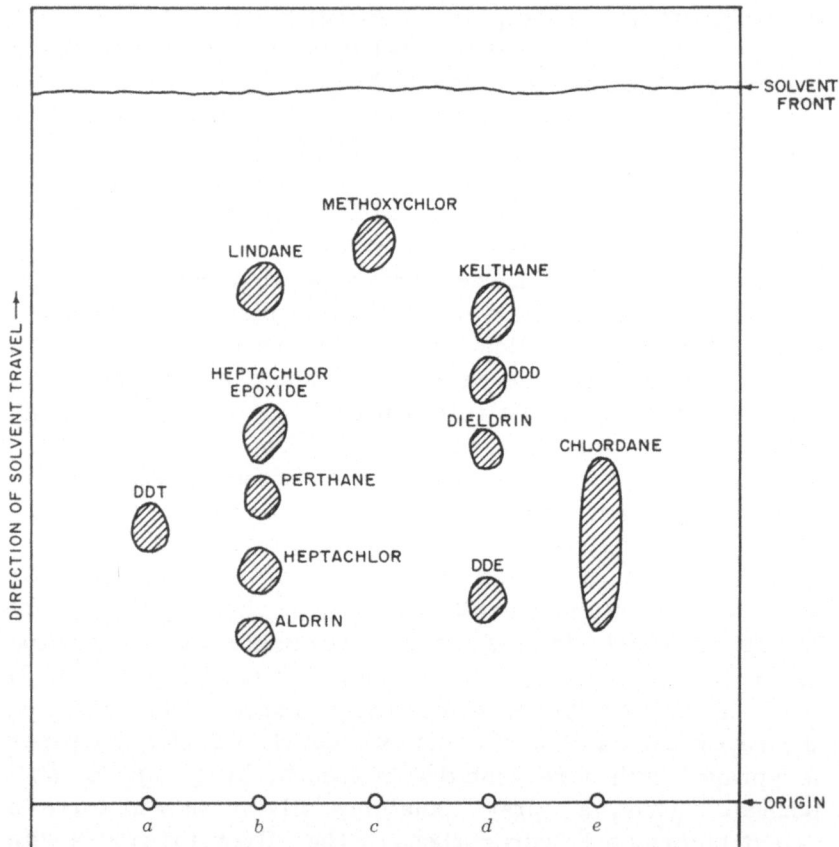

Fig. 1. Paper chromatogram showing: (a) migration of DDT, (b) separation of mixture of five pesticides, (c) migration of methoxychlor, (d) separation of mixture of four pesticides, and (e) streaking of chlordane.

Thin–Layer Chromatography

Thin-layer chromatography closely resembles paper chromatography except that the stationary phase used to retard the migration of the pesticides is an adsorbent such as silica gel or alumina rather than paper. The adsorbent (or active solid) is mixed with a binder and slurried in a solvent. The slurry is then deposited in a thin layer on a glass plate, and the surface is smoothed with a doctor blade. The plate is dried to remove the solvent. At the end of this process, the adsorbent is bound firmly in a thin layer to the surface of the glass plate. A sample containing pesticides is applied to the lower end of the plate and the solvent is permitted to evaporate. The bottom end of the plate is then immersed in a reservoir of solvent contained in a sealed cabinet. As in paper chromatography, the solvent migrates by capillarity along the thin layer of active solid, carrying the pesticides with it at different rates. When the solvent front approaches the top of the plate, the chamber is opened and the plate is removed and dried. The location of the spots on the plate is determined by spraying with specific reagents as described for paper chromatography. The advantages of thin-layer chromatography over paper chromatography are that it is rapid, the properties of the stationary phase can be varied over a wide range by the choice and pretreatment of the active solid, and it is more sensitive. Thin-layer chromatography is also a very useful supplementary procedure to clean up samples prior to gas chromatography.

Gas Chromatography

Gas chromatography is the most useful single technique presently available to the residue analyst. It differs from paper and thin-layer chromatography in that a permanent gas such as nitrogen, argon, or helium is used as the mobile phase. The stationary phase is prepared by impregnating an inert solid with a nonvolatile organic liquid. A free-flowing powder is formed that is then packed into a glass tube usually $\frac{1}{4}$ in. in diameter and 6 ft long. The tube is installed in a heated cabinet and a stream of gas is forced through it at a constant rate. A solution containing a mixture of pesticides is injected into the hot column at the end nearest the gas inlet. The pesticides vaporize and are carried through the column by the moving gas. They travel through the column at different rates, owing to differences in their partition coefficients between the

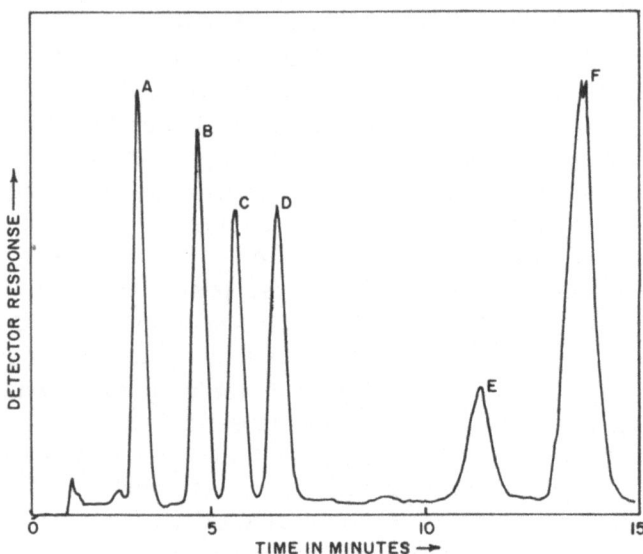

Fig. 2. Typical gas chromatogram obtained on a mixture of pesticide standards.

liquid stationary phase and the mobile gas phase. Consequently, they emerge from the far end of the column at different times and pass through a detector that automatically measures their concentrations. The response of the detector is fed into a strip-chart recorder so that the presence of each compound is registered as a peak and is measured as it emerges from the chromatograph. A typical gas chromatogram obtained on standard pesticides is shown in Fig. 2. The distance of each peak from the origin is known as the retention time and is characteristic for each pesticide. The area under the peak is proportional to the amount of compound passed through the detector. Thus, through the use of gas chromatography, it is possible to obtain both a qualitative and quantitative measure of the pesticides present in the sample.

The gas chromatographic method provides very high resolution combined with excellent sensitivity and specificity. Thus, complex mixtures containing lindane, heptachlor, aldrin, heptachlor epoxide, and DDT can be resolved in a single operation. As little as 10 picograms (pg) to 10 nanograms (ng) of each compound can be detected, sensitivity being dependent on the type of detector used, the compound, and the level of background noise.

Several detectors are available that are capable of indicating the presence of pesticides with very high sensitivity and selectivity. The most sensitive of these is the electron capture detector. It is capable of measuring some pesticides in amounts as low as 10 pg. This detector yields extremely high responses to halogen-containing compounds and some organic phosphates but is relatively insensitive to many compounds that do not contain halogen. This reduces background noise and makes it possible to detect peaks arising from pesticides, even when impurities are present in the sample. However, specificity is not absolute. Many oxygenated hydrocarbons, particularly those containing carbonyl groups, give high responses to electron capture. Thus, it is not always possible to discriminate between peaks that arise from small amounts of chlorinated hydrocarbons and those that arise from larger amounts of oxygenated compounds.

When the extreme sensitivity of the electron capture detector is not required, the use of the microcoulometric detector is preferred because of its high degree of selectivity and the negligible background produced by most impurities. The microcoulometer is capable of detecting halogen and sulfur with absolute specificity. Thus, when a peak is observed on the gas chromatogram, it is certain that the compound giving rise to it contains the element being monitored. This minimizes the chances of confusing peaks arising from impurities with those of pesticides. The microcoulometer operates on the following principles. As each compound emerges from the chromatographic column, it passes through a combustion tube where it is burned with oxygen to yield water and carbon dioxide. If the compound contains halogen and sulfur, hydrogen chloride and sulfur dioxide are formed also. The gas stream then flows through a titration cell containing silver ion, which is maintained at a constant concentration electrochemically. Hydrogen chloride, if present, precipitates the silver ion stoichiometrically, and the current required to regenerate it from the silver electrode is recorded as a gas chromatographic peak. Sulfur dioxide can be measured by replacing the chloride-sensitive titration cell with an oxidation-reduction cell. Using the model C-200 microcoulometer, as little as 10 ng of chlorine or sulfur can be measured. Hence, the procedure is more sensitive than any other currently available, with the exception of the electron capture detector.

SPECIAL PROBLEMS IN RESIDUE ANALYSIS

The methods described above are generally applicable to chlorinated hydrocarbon insecticides and allied compounds. However, a fairly large number of chemicals are now used in agriculture that do not fall into these categories. Some of these present special problems in analysis that require modification of the standard techniques used for chlorinated hydrocarbons.

Organic Phosphates

Generally applicable methods for the analysis of organic phosphates currently available are not very satisfactory. Extraction of the parent compounds from most media is no more difficult than in the case of chlorinated hydrocarbons. However, oxidation and hydrolysis products are formed rapidly on exposure of these compounds to the environment. These are more polar and water soluble than the parent compounds and, consequently, are more difficult to extract. Difficulty has been encountered in extracting many of these materials from soil, either due to decomposition or irreversible binding to soil particles. General procedures for cleaning up these extracts are also presently unavailable, although they are being worked on in many laboratories. Some phosphates decompose on or are bound irreversibly by the adsorbents used for column chromatography of chlorinated hydrocarbons. Several column methods have been developed specifically for them, but they have been applied to only a limited number of compounds.

After extraction and cleanup, paper chromatography is commonly used for separation and analysis. Adequate resolution is obtained, and it is possible to detect a number of com-pounds specifically through employment of the appropriate chemical reagents. Thin-layer chromatography is just coming into use for the analysis of organic phosphates, and the preliminary results available look very promising.

Gas chromatography also can be used for the measurement of phosphates. Column conditions have not yet been established that will permit the elution of all of the phosphates quantita-tively. However, recoveries are reasonably good in many cases. More difficulty is experienced in chromatographing some of the oxidation and hydrolysis products of phosphates. However, the chances are good that these can be converted to stable compounds by reduction and methylation.

Several detection systems are currently being evaluated for the selective detection of phosphates. In one of these, the organic phosphates are reduced to phosphine, and this gas is measured selectively with the microcoulometric titration sys-tem [2]. Another procedure makes use of a flame ionization detector in which a coat of a sodium salt is fused onto the electrode. This enhances the responses obtained to organic phosphates compared to those yielded by hydrocarbons and chlorinated hydrocarbons [3]. Finally, a method has been developed by McCormack et al. [4] in which the column effluent is analyzed by emission spectroscopy with the monochrometer set at the phosphorus line at 2539 Å.

Herbicides

Herbicides such as 2,4-D and 2,4,5-T are difficult to extract from most samples. Because they are anionic, the substrate usually must be acidified prior to extraction and a polar solvent used. These compounds can occur in a number of forms in the sample, so provision must be made for all of these in order to obtain quantitative results. Thus, esters of 2,4-D as well as salts are used in agriculture. These esters can hydrolyze to the free acid, particularly when taken up by plant tissues. Moreover, the free acids can become conjugated with sugars or other metabolites. These conjugates are impossible to extract by conventional procedures; therefore, they should be hydrolyzed prior to extraction. Many published methods do not provide for this. For extraction to be quantitative, it is necessary to provide for recovery of the free acid, esters, and possible conjugates with naturally occurring metabolites. Also, it has been found that proteins present in some samples can result in poor recoveries of 2,4-D. Evidently, they are capable of binding 2,4-D at cationic sites, thus shifting the partition coefficient in the direction of the aqueous phase. This effect can be eliminated by hydrolyzing the protein enzymatically prior to extraction.

Gas-liquid chromatography employing the microcoulometric or electron capture detector is the most satisfactory method currently available for the analysis of the compounds. The acids are usually converted to methyl esters by treatment of the residue with diazomethane or methanol-BF_3. Yields of the esters are satisfactory, and they chromatograph without decomposition.

IDENTIFICATION BY INFRARED ABSORPTION

Excellent circumstantial evidence leading to the identification of unknown pesticides can be obtained by gas chromatography or a combination of gas and thin-layer chromatography. However, further confirmation of identity by optical methods is generally required for absolute proof. Identification by infrared spectrophotometry is the most satisfactory technique currently available. Other methods are sometimes employed, but they have not received general acceptance. These include mass spectrometry, nuclear magnetic resonance spectrometry, ultraviolet spectrophotometry, and sometimes combinations of these. Some of these techniques lack the specificity required in pesticide residue analysis. Others require large samples and highly skilled personnel for operation and interpretation of results, are high in cost, or are not developed to the point where procedures can be recommended for routine laboratory use. Infrared spectrophotometry is therefore the most satisfactory method presently available. However, the limited quantity of purified pesticides recoverable from typical residue samples often presents considerable problems.

Therefore, the infrared spectrophotometer serves primarily as a supplement to the more sensitive quantitative instruments such as the gas chromatograph. It provides evidence by means of recorded spectra which, like fingerprints, are used for unequivocal identification of compounds. The spectrum of an unknown is compared with those of standards; identity is established if one of the traces is superimposable with that of the unknown. Even when complete matching of all details of the traces is not achieved, it is possible to identify or eliminate candidate functional groups, since most groups possess characteristic absorption bands.

The application of infrared spectrophotometry to residue analysis is dependent on obtaining a highly purified sample by chromatography. Because the most severe limitation up to now has been lack of sensitivity of the infrared spectrophotometer, much less has been published on residue analysis than on analysis of formulations where much larger quantities are available. However, improvements in instrumentation and refinements in sample isolation techniques are obviously brightening the outlook for fuller exploitation of infrared techniques in residue work.

Samples of unknown pesticides are isolated by gas or thin-layer chromatography and prepared for infrared analysis by incorporating them in KBr microdiscs or placing them in microcavity cells, with or without an organic solvent. The spectra are then obtained in the usual way and compared with the spectra of known pesticides in order to obtain the best matching pair.

For best matching, spectra of both unknowns and standards should be obtained on the same type of instrument, preferably on the same instrument. In this way, the spectra can be compared directly by superimposing the traces. The spectrum of the insecticide dimethoate is shown in Fig. 3.

Many factors contribute to the type of spectrum obtained, sample concentration being one of these. If the sample is too dilute, only the largest peaks will be seen, while if the sample

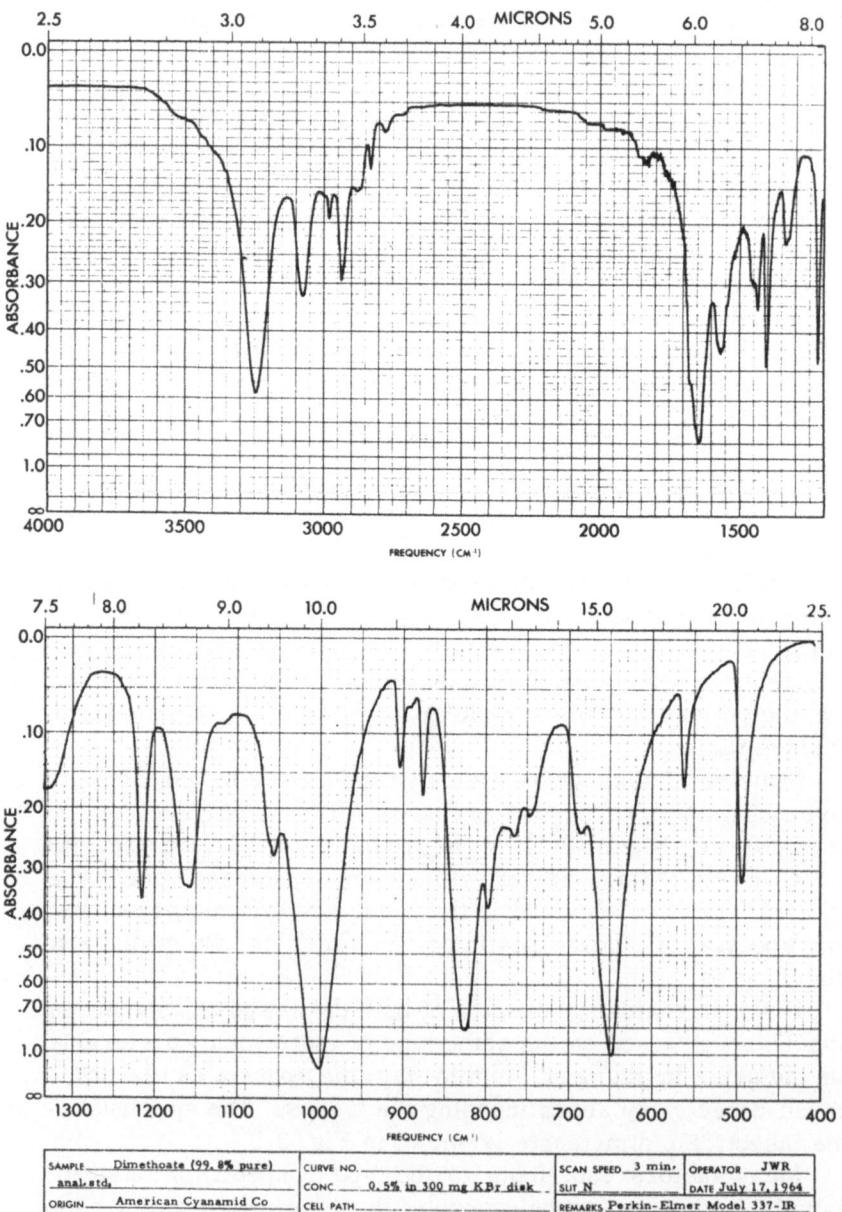

Fig. 3

is too concentrated, larger peaks will be off-scale. In either case, it would be hard to compare the spectra with those of known compounds, and important spectral information would be lost. Thus, it is often necessary to run a trial spectrum and then adjust the sample size, solution concentration, or film thickness according to the results obtained.

Another important factor when comparing spectra is to present the sample and standard in the same way. Thus, an unknown sample prepared in a KBr disc should be compared to known standards with spectra obtained in KBr discs and films compared with films. Solutions should always be prepared in the same solvents, and spectra of these solvents should also be obtained in cells of the same size so that the analyst is aware of any possible peak interferences. In brief, the unknown and the standard compounds should be handled by techniques as similar as possible to obtain the best correlation of spectra. Complete records of the method of preparation and presentation of the sample should be included on the spectra of standard compounds and unknown samples.

Despite all possible care in the preparation of unknowns and standards, the spectra may not superimpose exactly. This can also occur if uniform sample preparation is not possible. However, comparisons of certain absorption peaks may lead to positive identification. The most important peaks are the largest peaks arising from absorption frequencies of functional groups, so that these should be examined with particular care. When the spectra are presented in different forms it is best to make a list of the frequencies of the major peaks. One system is to record in decreasing order of peak size the frequencies of the largest peaks of the unknown sample. The frequencies of the comparable peaks on the standard spectrum are then placed directly opposite. Correspondence of the major peaks with respect to size, order, and frequency constitute positive identification. It is generally agreed that at least the three major peaks must agree before proof of identity is acceptable. These three peaks should not arise from the same functional group, such as $C=O$, $P-O$, or $C-Cl$. Also, these major peaks should not be those arising from C-H vibrations. C-H stretching frequencies occur around 3000 to 2700 cm^{-1}, while

C-H deformation frequencies occur around 1450 to 1350 cm^{-1}. Thus, major peaks in these regions should not be considered.

When spectra are not completely superimposable and the method of peak comparison is used, it is advisable for the analyst to have some knowledge of spectral interpretation in regions of functional group absorption. Absorption maxima in different regions of the infrared indicate different types of vibrations between different atoms in the molecule. These regions, and types of absorption bands produced, are specific enough from molecule to molecule so that much can be learned about unknown samples without references to standard compounds. Often the molecule, by the spectrum alone, can be positively identified by a spectroscopist well trained in this field. Thus, organic phosphates have a characteristic absorption band dependent upon vibration between the phosphorus atom and other atoms or groups. The P-O (free) vibration produces a strong absorption maximum in the region around 1350 to 1250 cm^{-1}, while P-O (hydrogen bonded) produces a strong maximum from 1250 to 1150 cm^{-1}. These are just two examples of many bands that might occur, depending upon the groups attached to phosphorus atoms and the molecular interactions produced. All of the major peaks of a spectrum must be studied and interpreted, and interrelations must be noted before an identification can be made. When the analyst has some knowledge of these absorption bands and the interactions that may cause them, he is in a much better position to interpret any doubtful spectra intelligently.

REFERENCES

1. H. P. Burchfield, D. E. Johnson, and E. E. Storks, "Guide to the Analysis of Pesticide Residues," (prepared for the U.S. Department of Health, Education, and Welfare, Public Health Service, Bureau of State Services, Environmental Health, Office of Pesticides). U.S. Government Printing Office, No. 26/2 — P43/2 Vols. I and II, 1965.
2. H. P. Burchfield, J. W. Rhoades, and R. J. Wheeler, "Simultaneous and Selective Detection of Phosphorus, Sulfur, and Halogen in Pesticides by Microcoulometric Gas Chromatography," J. Argic. Food Chem. 13:511, 1965.
3. L. Giuffrida, "A Flame Ionization Detector Highly Selective and Sensitive to Phosphorus—A Sodium Thermionic Detector," J. Assoc. Off. Agr. Chem. 47:293, 1964.
4. A. N. McCormack, S. C. Tong, and W. D. Cooke, "Sensitive Selective Gas Chromatography Detector Based on Emission Spectrometry of Organic Compounds," Anal. Chem. 37:1470, 1965.

A Modified Hydrogen Flame-Ionization Detector, Highly Sensitive and Selective for Phosphates

R. V. Hoffman

Food and Drug Administration
Buffalo, New York

INTRODUCTION

Microcoulometric and electron capture detectors have been routinely used for many years in pesticide residue analysis. With the microcoulometric systems one is capable of detecting halogens, sulfur, and phosphorus with high specificity but with somewhat low sensitivity, while with the electron capture detector one can detect compounds containing strong electronegative groups with high sensitivity but with low specificity. While in most cases the electron capture mode of detection is relatively sensitive to most organic phosphate pesticides, its lack of selectivity tends to limit the usefulness of this detector. It was fortunate that a detector was developed for these organophosphate pesticides. This detector was the sodium thermionic detector.

Giuffrida [1] developed a sodium sulfate modified hydrogen flame ionization detector that was referred to by Giuffrida as the sodium thermionic detector. Karmen [2] also made contributions to the development of this mode of detection. This detector is highly sensitive and specific for organophosphate

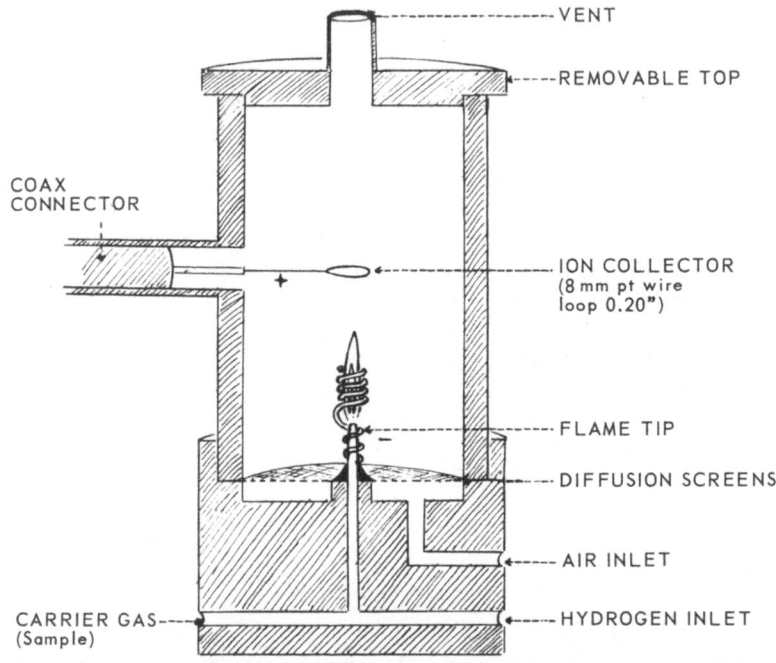

Fig. 1. Commercial flame ionization detector.

pesticides, and yet it may be adjusted to show only a limited sensitivity for chlorinated pesticides. In this paper, Giuffrida [1] discussed the construction of the detector, the operational parameters, and the sensitivity of the detector for several organophosphate pesticides.

In subsequent papers, Giuffrida et al. [3, 4] discussed further work with the thermionic detector. The latter paper [4] is of particular interest because it describes the responses to phosphorus obtained with alkali metals other than sodium, and it also describes a procedure for incorporating the thermionic detector into a dual detection system with an electron capture detector.

The following discussion will elaborate on Giuffrida's work on the thermionic detector; however, the emphasis will be placed on the actual mechanics of modifying the flame ionization cell, the various operational parameters, and,

finally, the results of an evaluation study of the KCl thermionic detector conducted by the Food and Drug Administration.

DISCUSSION

General

Figure 1 shows a typical flame ionization detection cell, which basically consists of a cell with two polarized electrodes, electrically insulated from each other, and a flame in which ionization is produced.

A 300 V battery is used to impress a potential across the cell electrodes. When the cell is at equilibrium, with the flame burning and the effluent gas flowing through the cell, the conductivity is a constant value. The battery is ordinarily connected in such a way as to make the platinum loop the anode; however, excellent results have been obtained by reversing the polarity. The small amount of current flowing through the cell is called the standing or base line current. When an organic compound is burned in the flame, a few of the molecules become ionized, thus increasing the base line current. This increase in base line current is directly proportional to the number of ionized carbon atoms generated by the flame, which in turn varies directly with the number of carbon atoms in a given compound. In the hydrogen flame ionization detector, according to Sternberg [5], the ionization efficiency is about 0.001% for carbon and hydrogen compounds and even less for those containing oxygen or halogens. By introducing an alkali metal salt electrode in the flame an additional source of ions is present that does not increase the ionization efficiency of the hydrocarbons, but it does increase the ionization efficiency of the organophosphates and halogens, thereby yielding greater selectivity for determining these compounds. Organic thiophosphates also show an enhanced response.

SELECTIVITY AND SENSITIVITY OF VARIOUS ALKALI METAL SALTS

Giuffrida [4] states that certain potassium, cesium, and rubidium salts of the highest state of purity are generally superior as coatings on the electrode to Na_2SO_4, which was originally used as the coating. In general, all potassium salts that can be coated on the electrode show the greatest response to phosphates; however, the halide salts KCl, KRr, or KI show the greatest selectivity, as these salts suppress the detector response to halides. Some rubidium (Rb) salts also suppress the halogen response.

The Na_2SO_4 thermionic detector enhances the response to phosphates about 600 times and 20 times for halides as compared with the response of a conventional hydrogen flame ionization detector. Highly purified KCl (reagent grade recrystallized at least twice in water) gives a response of about 3 times that of reagent grade Na_2SO_4 for the phosphates, while suppressing the response to halides. It is possible to detect 1 nanogram (ng) of parathion with either detector coating.

As with any flame detector, one cannot use aromatic or halogen solvents with the thermionic detector. These solvents undergo incomplete thermal decomposition, thereby leaving carbon deposits. This may result in high base line currents and/or uneven detector response.

There are a number of disadvantages connected with utilizing KCl as a coating. When this detector is in use a fine deposit of KCl forms within the detector and results in decreased sensitivity after a period of 2 months. This necessitates cleaning of the detector and recoating of the electrode. In addition, initial equilibration of the detector takes considerably longer with the KCl coating than with Na_2SO_4. Generally in speaking, in spite of its disadvantages, KCl is to be preferred because of its greater sensitivity.

PREPARATION AND COATING OF ELECTRODE OR HELIX

A critical parameter for obtaining optimum performance from the thermionic detector is the construction of the helix to which is fused the alkali metal salt.

Figure 2 shows a drawing of a coiled, coated electrode or helix installed in the proper position on a flame jet. The helix

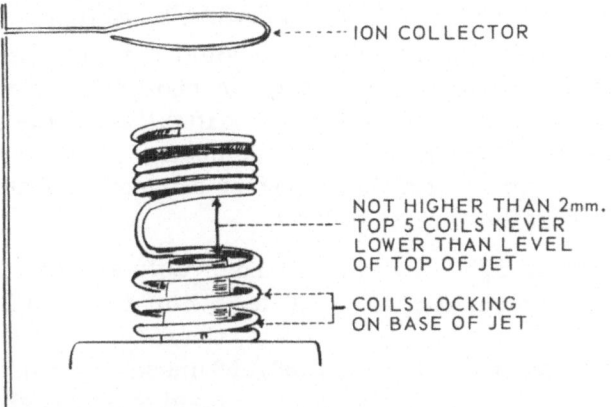

Fig. 2. Helix installed on flame jet.

itself consists of a piece of platinum or platinum-iridium wire, 0.016 to 0.019 in., coiled in such a manner as to have a number of turns that anchor to the flame jet and about five coils suspended above the flame jet. The coils above the jet should be approximately 5 mm in diameter and, for this purpose, a 10/32 machine screw may be used as a mandrel. After the top coils are formed, they should be compressed together as close as possible. The three or more turns that anchor onto the flame jet should be made to fit the jet in a fairly snug manner. An appropriate sized mandrel should be used for this purpose. The helix should be positioned on the jet so that the bottom of the five suspended coils is not lower than the top of the jet [4] nor higher than 2 mm above the jet. Maximum ionization efficiency, and therefore sensitivity to phosphates, depends directly upon the proper heating of the alkali metal salt in the flame. Incorrect positioning of the helix has resulted in as much as 60% loss in sensitivity to phosphorus.

After constructing the helix, it must be fused with the alkali metal salt. This is best done by applying a few drops of a saturated solution of the salt to the helix and then heating it to red heat in a bunsen burner. As the salt flows over the helix, the helix should be turned slowly in the flame. Repeating this procedure 3 or 4 times insures a good covering of the helix with the fused salt.

OPERATIONAL PARAMETERS

The conditions necessary to yield optimum performance for gas chromatography of organic phosphate pesticides will naturally depend on many factors that will effect the separation and response to any particular compound. However, the conditions for gas chromatography of organophosphates we use in the laboratory are as follows:

Chromatograph . . . . .Barber–Colman 5000

Column dimensions . . . . . .6 ft, 6 mm OD, 4 mm ID glass U–tube

Solid support . . . . .80/100 mesh Gas Chrom Q (Applied Science Laboratories)

Stationary phase . . . . . . .10% DC 200 (Applied Science Laboratories)

Temperatures . . . . .injection, 240°; column, 205°; detector, 190°

Carrier gas . . . . .nitrogen at 60 ml/min

Combustion gases . . . . . . .hydrogen at 38 ml/min (Na_2SO_4), 28 ml/min (KCl); air at 300 to 325 ml/min

Detector . . . . .Na_2SO_4 or KCl modified hydrogen flame ionization detector "thermionic"; cell voltage: 300 V

Sample size . . . . .1.0 to 2.0 ng per μl in iso–octane (2,2,4-trimethylpentane), or if not soluble, in acetone

EVALUATION STUDY

During the summer of 1965 the Food and Drug Administration conducted, in several of its district laboratories, an evaluation study of the KCl modified hydrogen flame ionization detector. The following parameters were determined in order to calibrate the detector and instrument:

Carbon Response of Flame Ionization Cell

Hydrogen . . . . .35 ml/min; sensitivity: 10^{-9} AFS (amperes full scale)

Test solution . . . . .1 μg parathion and DDT per 5 μl
                                 iso-octane
Amount injected . . . . .5 μl
Response. . . . . .Parathion–30%; deflection DDT–
                                 minimum detectable re-
                                 sponse

and other instrumental parameters, as stated previously.

Response of Thermionic Detectors

	Na_2SO_4	KCl
Hydrogen flow	. . .35 ml/min	25 ml/min
Sensitivity	. . . . .10^{-8} AFS	3 · 10^{-8} AFS
Test solution	. . . .10 ng parathion and DDT/5 μl iso-octane	same
Response	. . . . . .parathion—42% deflection DDT—no response	48% deflection no response

and other instrumental parameters, as stated previously.

Table I.   Relative Sensitivities to Parathion[a]

Pesticide	%P[b]	R% P[c]	Participating lab			
			1	2[d]	3	4
Thimet	13.5	1.27	1.1	1.06	0.74	1.53
Diazinon	10.16	0.96	1.1	1.03	0.92	0.65
Methyl parathion	11.7	1.10	1.1	1.02	0.75	0.40
Ronnel	12.3	1.16	1.0	0.96	0.92	0.61
Malathion	9.4	0.89	0.7	0.63	0.22	0.29
Parathion	10.6	1.00	1.0	1.00	1.00	1.00
Ethion	16.1	1.52	1.4	1.32	1.06	0.95
Trithion	9.0	0.85	0.8	0.77	0.59	0.64
Guthion	9.8	0.92	0.4	0.35	—	0.09
Absolute response of parathion in mm^2 per nanogram			31	29	27	35

[a] Relative sensitivity = response of organophosphate/response of parathion.
[b] % P = phosphorus in pesticide molecule.
[c] R % P = relative percent phosphorus in molecule to parathion.
[d] Lab 2 = Buffalo Laboratory.

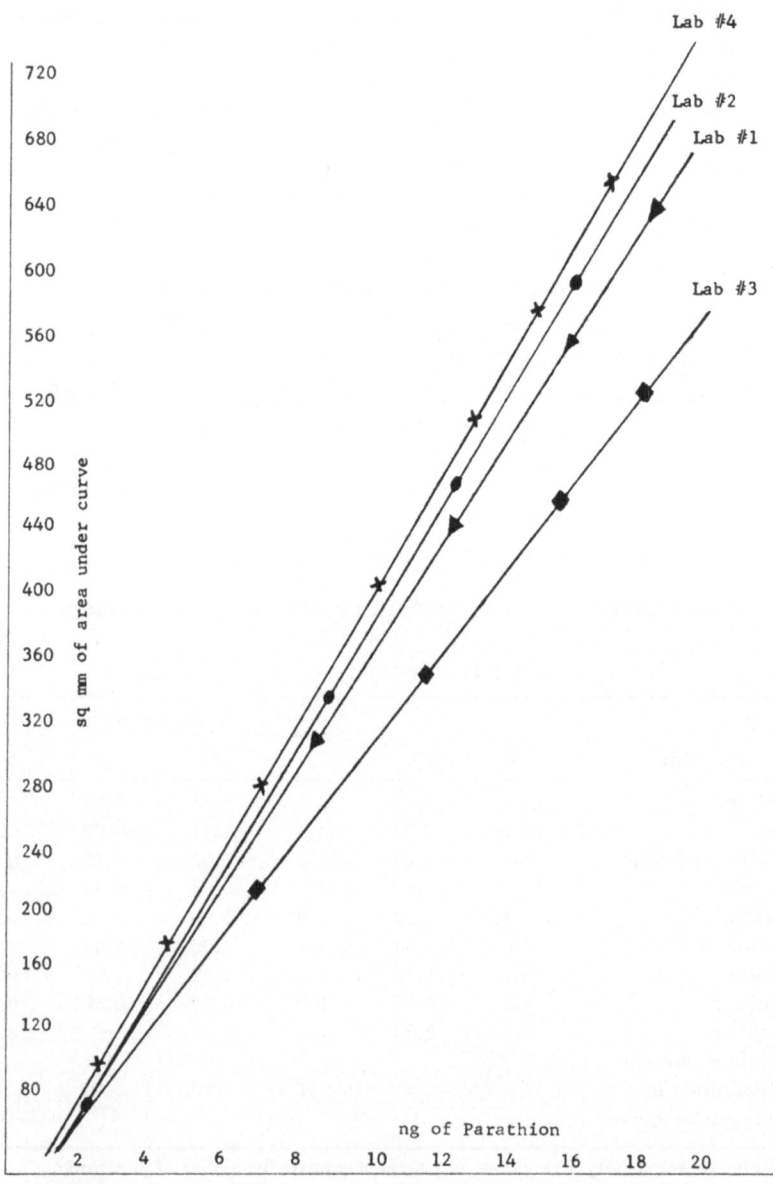

Fig. 3. Thermionic detector response to parathion (potassium salt).

The results of the evaluation study of the KCl thermionic detector are as follows:

Relative Sensitivities to Parathion

Table I indicates that, generally speaking, the detector response is directly proportional to the percent of phosphorus in the pesticide molecule. The cases where this is not borne out are probably due to experimental difficulties with the instrumental and column parameters.

Relative Retention Compared to Parathion

The data in Table II indicate that there is no difficulty in identifying the various pesticides studied by noting their retention times relative to parathion.

Linearity of Detector Response

The evaluation study indicated that the detector response was linear up to 20 ng for the pesticides studied. Linearity was determined by making five injections at each 2-ng level and then calculating the average area for each level of concentration. Concentration was then plotted versus average area in order to determine the linearity curve. Figure 3 shows a

Table II. Relative Retention Compared to Parathion[a]

Pesticide	Lab 1	Lab 2	Lab 3	Lab 4
Thimet	0.38	0.37	0.41	0.43
Diazinon	0.52	0.52	0.54	0.53
Methyl parathion	0.71	0.71	0.75	0.71
Ronnel	0.82	0.82	0.81	0.83
Malathion	0.91	0.92	0.91	0.89
Parathion	1.00	1.00	1.00	1.00
Ethion	2.54	2.57	2.31	2.42
Trithion	2.93	2.94	2.96	2.78
Guthion	5.18	5.22	—	4.80

[a] Relative retention = retention of organophosphate/retention of parathion.

Table III.   Results of Analysis of Unknown Solutions of Organophosphate Pesticides, Laboratory 2

Solution	Pesticide	Concentration added	Results	
			Trial 1	Trial 2
C	Parathion	1.07	1.05	1.07
	Thimet	1.11	1.10	1.10
	Ronnel	1.00	0.95	0.99
CC	Parathion	1.05	1.11	1.10
	Thimet	1.11	1.17	1.15
	Ronnel	1.03	1.10	1.01
D	Parathion	2.14	2.12	2.11
	Thimet	2.22	2.06	2.15
	Ronnel	2.00	1.82	1.86
DD	Parathion	2.10	2.15	2.13
	Thimet	2.22	2.12	2.11
	Ronnel	2.06	2.02	2.03

curve for parathion. It should be noted that Laboratory No. 2 has reported that by carefully controlling the operational parameters, a linearity for most of the studied pesticides up to 35 ng can be achieved.

Identification and Quantitation of Unknown Organophosphate Residues

Table III shows Laboratory No. 2's qualitative and quantitative results of four different unknown solutions. The results proved to be satisfactory in that accurate identification and quantitation of these unknown pesticides were achieved.

CONCLUSION

Although the theoretical basis for the enhanced response of the hydrogen flame ionization detector when an alkali metal-coated electrode is used is unknown, this detector can still be utilized for effective screening of organophosphate pesticides at nanogram levels with no appreciable base line noise. Both the sodium and potassium thermionic detectors have their own specific advantages: the sodium detector maintains its sensitivity over a longer period of time, while the potassium detector has greater sensitivity. Both detectors show linear response up to 35 ng for most common organophosphate pesticides. Work is continuing in order to improve the sensitivity, specificity, and stability of the detection system.

REFERENCES

1. L. Giuffrida, "A Flame Ionization Detector Highly Selective and Sensitive to Phosphorus—A Sodium Thermionic Detector," J. Assoc. Off. Agr. Chem. 47:293, 1964.
2. A Karmen, Anal. Chem. 36:1416, 1964.
3. L. Giuffrida and N. F. Ives, J. Assoc. Off. Agr. Chem. 47:1112, 1964.
4. L. Giuffrida, N. F. Ives, and D.C. Bostwick, J. Assoc. Off. Anal. Chem. 49:8, 1966.
5. J.C. Sternberg, W.S. Gallaway, and T.L. Jones, "The Mechanism of Response of Flame Ionization Detectors," in: Gas Chromatography, Third International Symposium, Academic Press, New York, 1962.

Galvanic-Coulometric Detectors in Gas Chromatography

Paul Hersch

Gould-National Batteries, Inc.
Minneapolis, Minnesota

INTRODUCTION

Among the most sensitive detectors for gas chromatography are those based on coulometry. They owe their sensitivity to the high value of the faraday and to the ease with which minute electric currents are measurable. Electrometers with a discrimination for less than 10^{-15} amp are available, a current corresponding to about 10^{-20} equivalents per sec, or 6000 univalent ions per sec. For example, take anthracene, $C_{14}H_{10}$. A coulometrically quantitative oxidation of one molecule means stripping it of $14 \times 4 + 10 \times 1 = 66$ electrons. Thus, a stream of fewer than 100 molecules per sec of this compound is detectable in principle. In the case of compounds with more carbon and hydrogen atoms in the molecule, the sensitivity in these terms is higher still, compensating for their lesser volatility.

For comparison, consider flame ionization detectors. They translate only a minute fraction of the carbon atoms into electric current, about 1 in 10^5. Coulometry, at least in principle, utilizes every carbon and hydrogen atom, yielding a sensitivity potentially superior to flame ionization by several orders of magnitude.

Despite this advantage, the presently available coulometric detectors have found use only in limited areas, such as insecticide analysis. The equipment is rather complex. Thus, organic chlorine, after pyrolysis of the compound to HCl, is carried to a continuous electrolytic titrator generating silver ions from a silver anode. The titrator comprises a potentiometric sensing circuit in addition to the reagent generating circuit. The former produces an error signal that corrects, after amplification and via a servo loop, the rate of generation, keeping it in step with the rate of arrival of HCl. The generating current is proportional to the flux and is recorded. Similarly, organic sulfur is converted to SO_2 by combustion, and the SO_2 is titrated continuously with bromine generated anodically from bromide. Again, a pair of potentiometric redox electrodes with amplifier and servo feedback are called in to ensure that neither more nor less bromine is produced in any differential of time than is needed for the oxidation of the SO_2 arriving in that differential. For nonspecific detection of organic compounds generally, they may be combusted, and the resulting CO_2 may be titrated with cathodically generated alkali on the same lines.

This chapter presents and recommends a simpler type of coulometric detector. In general, the analyte in the effluent is made the subject of some chemistry—pyrolysis or combustion—as in the above methods, or milder reactions. The gaseous reaction product (e.g., I_2, HCl, or SO_2) or the unconsumed residue of a gaseous reagent (e.g., O_2, ˙Br_2) is then carried to a cell to actuate a galvanic system, with quantitative coulombic yield. "Galvanic system" means an assembly of an anode, electrolyte, and cathode that generates current without the aid of an external electromotive source. Amperometric systems, that is, with an external driving force, are also applicable and can also result in fully coulometric yields, but generally they pass a substantial background current in the absence of an electromotively active species, while galvanic systems are more nearly self-zeroing.

Basically, the new detectors have only two, not four, electrodes, and their circuitry consists of nothing more than a pair of wires leading to a galvanometric recorder or to a resistor in parallel with a potentiometric recorder. There is no need for servo mechanisms nor for any moving parts such as stirrers. Such simplicity permits lower costs and makes malfunctioning less probable.

HISTORY

A galvanic method for the detection of organic compounds in gas chromatography was proposed by this writer in 1959 [14]. Before that time, he had developed a galvanic, though non-coulometric, cell, convenient for monitoring oxygen in parts per million ranges [13]. It was based on the system

Gas/Ag screen/porous PVC sheet + aq KOH/Pb foil

The current generating process was

At the Ag cathode: $\frac{1}{2}O_2 + H_2O + 2e \rightarrow 2OH^-$
At the Pb anode: $Pb + 3OH^- - 2e \rightarrow PbO_2H^- + H_2O$

Only a proportion—about one in ten—of the oxygen entering this cell was cathodically absorbed and translated to current; the rest of the oxygen escaped. Because the proportion utilized, the coulometric yield y, varied from cell to cell, as well as with age and temperature, frequent standardization with electrolytically generated traces of oxygen was essential.

In applying this cell to gas chromatography, a trace of air is bled into the effluent. The gas stream passes then over a hot catalyst, to combust the organic transients, and then through the sensor. The air bleed establishes a constant base line of galvanic output, and each combustible registers as a negative peak, or valley, as in the electron capture method. If the nature of the compound is known, simple stoichiometry permits the calculation of its weight from the area of the valley, taking into account the coulometric yield, but without the need for a standard sample. Taking again anthracene as an example (equivalent weight = 78/66 = 1.18), a valley area of $1\,\mu$amp-sec represents

$$\frac{1}{y} \cdot \frac{10^{-6}}{96,489} \cdot 1.18 = 1.22 \cdot \frac{10^{-11}}{y} \text{ gram}$$

In general, no electrometer-amplifier is needed.

More recently, T. R. Phillips et al. [27] applied this cell to the determination of gas chromatographic transients of oxygen itself, in mixtures of permanent gases, making it possible to employ molecular sieve columns at room temperature instead of −70°C, with the benefit of greater speed and convenience. In the case of plant streams and reaction mixtures where oxygen was the only permanent gas of interest, it was no longer necessary to separate the components of the inert gas peak. Similarly, Hillman and Lightwood [23] found the cell to be of service in the gas chromatography of fuel gases where normally, with conventional detectors, the determination of oxygen suffers from interference by argon.

In 1959 and subsequently, A. Berton [6, 7] described several galvanic systems responsive to traces of a large variety of gases and vapors in air, using batch sampling. In one type of cell ionizable gaseous species actuated the system by creating conductivity rather than acting as depolarizers. In other types of cells containing strongly oxidizing electrolytes such as chromic acid between inert electrodes, reducing vapors (e.g., of alcohols, and even paraffins) seem to have established some kind of concentration cell. The coulometric yield appears to have been extremely low ($y < 0.001$). Also, the yield fluctuated, requiring *ad hoc* standardization with a known quantity of the species of interest before or after each determination. Despite these weaknesses, Berton's cells produced usable gas chromatograms of mixtures of alcohols and, with the aid of pyrolysis, of chlorinated hydrocarbons. He also "fingerprinted" town gas, automobile exhaust, cigarette fumes, pyrolysates of fatty and polymeric materials, and emissions from pine trees.

It is the main objective of this article to draw attention to a variety of possible future uses of galvanic detectors that also offer, besides simplicity, the advantages of coulometry. While a good measure of experience with the galvanic-coulometric sensors themselves, as continuous monitors, has been laid down in the literature [15, 16, 18–22], their application to gas chromatography has as yet been limited to no more than sketchy preliminaries. Present assignments in another field exclude this writer from this territory, except for this outline. He hopes that some of his anticipations and speculations will attract readers to further explorations.

CELL TYPES AND CAPABILITIES

The arsenal of galvanic-coulometric sensors that may be recommended for gas chromatography at this time is as yet limited to essentially two types, responsive to (1) molecular oxygen in inert gases and (2) halogen, ozone, vapors of strong acids, sulfur dioxide, and some other reducents in air and other gases.

First, a survey shall be made of the capabilities of each type in gas chromatography. A detailed description of the sensors themselves follows later.

Detection in Terms of Oxygen

The galvanic-coulometric cell sensitive to oxygen is directly applicable in those instances in which the galvanic non-coulometric cell has been used in the past, such as isolating the oxygen peak in the presence of argon [23, 27]. More importantly, the new sensor provides a general, nonspecific detector for organic eluates, if one applies the same decrement method as was proposed in 1959 when only the less-sensitive, noncoulometric cell was known. The advantage of direct quantitative interpretability of the valley areas without reference to standards is, of course, retained. In addition, the new cell offers higher sensitivity still and dispenses with the need for frequent calibration. A low-noise oxygen base line is required, but this is no serious problem. For the combustion step, equipment from elemental organic microanalysis is available, although it would be desirable to develop simpler, less bulky heating means to match the simplicity of the sensor itself. The system may be represented thus:

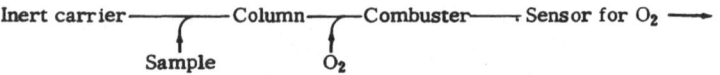

The oxygen can be added by means of electrolysis or through an air bleed, as described below. A scroll of fine platinum screen in a ceramic tube aids the combustion. The temperature should be at least 1000°C, and it should be kept constant. Platinum itself consumes some oxygen, presumably forming a volatile oxide. Fluctuations of temperature may affect the rate of this process and thus the base line.

The detection method in terms of oxygen consumed in combustion has one limitation, common to all decrement methods. This is so when a major constituent A, emerging from the gas chromatography column, is followed by a minor constituent B that should not be lost for analysis. Any oxygen base level sufficient for the complete combustion of A experiences only a relatively small depression from B, so that the accuracy in determining B is poor (Fig. 1a). On the other hand, if the oxygen level is adjusted down to become comparable in magnitude with B, and the requirement of A is ignored, then A pyrolyzes and carburizes, forming a carbon deposit in the combustion chamber. This deposit continues to consume oxygen with the result that the signal for B is completely swamped (Fig. 1b).

As a remedy to this situation, it is suggested that the oxygen base line be tailored to the needs of B and that moisture be added to the gas stream as an auxiliary oxidant for A. During the A-transient, all oxygen is used up, and that portion of A that is short of oxygen, instead of pyrolyzing and depositing carbon, reacts with water vapor forming H_2 and CO, gases that do not affect the sensor. This sacrifices the determination of A but makes the resolution and accurate analysis of B practicable (Fig. 1c). The compound A may still be deter-

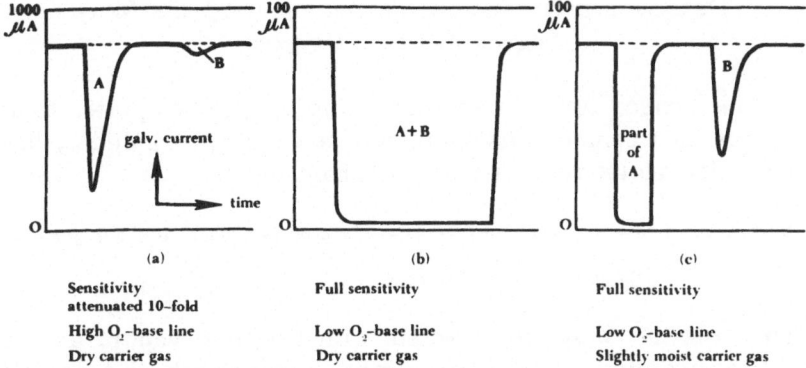

Fig. 1. O_2-decrement method of detection without and with H_2O as an auxiliary oxidant.

mined simultaneously by splitting the effluent and passing the branched-off stream through a second train, adding enough O_2 for the combustion of *A*, and using a separate, attenuated coulometric sensor. Also, *A* may be determined in the secondary stream by conventional, less sensitive means.

The addition of moisture is automatic if the addition of O_2 is by anodic evolution from aqueous electrolyte. Only very low relative humidities are required. For example, a relative humidity of 1% at 20°C in a carrier gas stream of 50 ml/min provides an "oxidant flux" of 1540 amp, sufficient to prevent carburization in most cases.

The technique of combustion above 1000°C, aided by water vapor, is meant to be applied, in the first place, to hydrocarbons and C–H–O compounds. Where other elements are involved, the stoichiometry of oxidation under the prevailing conditions—high dilution of both substrate and oxidant, and limited residence time in the reactor—may not be clear-cut. Even the nature of the products may depend on the operating factors. With some species, temperatures considerably higher than 1000° may be needed to ensure interpretability in absolute terms, or at least reproducibility.

Detection in Terms of Halogen and Oxidants More Aggressive than Oxygen

The sensor for halogen, in principle, may be applied to all species that consume halogen, spontaneously, or aided by heat or by ultraviolet irradiation. In "shorthand":

Inert carrier ————┬——Column————┬——Reactor——Sensor for halogen ——➤
　　　　　　　　　│　　　　　　　│
　　　　　　　Sample　　　　　Halogen

Thus, olefins may be reacted with bromine vapor, and the excess bromine may be monitored galvanically. Austin [3] has described an olefin monitor based on bromine addition but using a potentiometric detector. At sufficiently high temperatures, chlorine could be made to abstract hydrogen from ammonia, amines, and many other compounds, even paraffins.

Addition of chlorine at a constant rate is best done by electrolysis, while for bromine and iodine evaporative pick-up methods are suited. These "doping" techniques shall be discussed below.

The halogen sensor responds also to other strongly oxidizing vapors such as tertiary butyl hypochlorite and, with the right cell configuration and electrolyte, to ozone. The former can be introduced conveniently through pick-up from liquid t-BuOCl (b.pt. 79°C). It should yield decrement signals in response to numerous organic compounds. Presently, little is known about gas phase chlorinations and dehydrochlorinations with this compound. The study of these reactions could open interesting analytical possibilities.

Decrement analysis with added ozone should give selective signals from compounds susceptible to ozonolysis and to oxidation by ozone under mild conditions. The ozone may be generated by ultraviolet light in an air stream, tributary to the carrier gas, joining it at the exit point of the column:

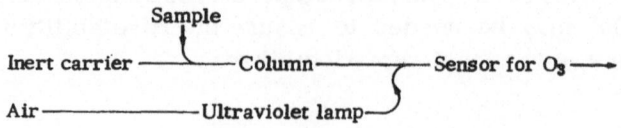

Sample
Inert carrier ————┴——Column————————Sensor for O_3 ——➤
Air ———————————Ultraviolet lamp—

Using a short column of iodine pentoxide kept at about 180°C, preceding an iodine sensor, a positive peak is attained from a transient of CO, owing to the century-old Ditte reaction:

$$5CO + I_2O_5 \longrightarrow 5CO_2 + I_2$$

A portable monitor based on this detection system has been constructed by this writer [16] and was used to measure carbon monoxide generated in freeway traffic situations [2, 12]. Ethylene and acetylene, like carbon monoxide, also react quantitatively with iodine pentoxide under suitable conditions [28]; other species, especially alcohols and carbonyl compounds, may react at least reproducibly. Thus, one may add the following scheme to the collection:

Carrier gas ————┬———— Column ———— I₂O₅ ————Sensor for I₂ ——▶
                                   Sample

Water, carbon dioxide, and oxygen-bearing organic compounds in an inert carrier gas stream may be cracked over incandescent carbon, converting them to CO and thus becoming amenable to galvanic-coulometric detection:

Nitrogen ————┬———— Column——Hot carbon——I₂O₅ —— Sensor for I₂ ——▶
Sample with O-compounds

Moreover, nonspecific detection of organics generally should be possible by oxidation over copper oxide, followed by reduction of the resulting CO_2 and H_2O by hot carbon:

Nitrogen ————┬——Column——CuO—— Hot C——I₂O₅ ——Sensor for I₂ ——▶
                         Sample

Each C atom from the sample yields 2 molecules of CO. This doubling effect may be repeated and the sensitivity, may be escalated as in a photomultiplier:

Nitrogen ————┬——Column ——CuO —— Hot C┐
                Sample   └CuO——Hot C┐
                                      └CuO——Hot C——I₂O₅——Sensor for I₂ ——▶

With K cycles CuO/hot C, 1 mole of a hydrocarbon $C_m H_{2n}$ should produce $0.4[2^k(m+n)-n]$ faraday. The several CuO beds could be accommodated in one common heater and the carbon beds in another.

Iodine pentoxide surely is not the only solid compound with positively valent halogen that can actuate a halogen sensor in response to specific analytes. Solid organic N-halogen compounds (chloramines and related bleaching agents) may be brought to react with weakly acidic eluents, evolving halogen. Furthermore, it may be fruitful to study the interaction of sodium chlorite with SO_2, H_2S, NO, and organic vapors. With SO_2, chlorine dioxide is known to be evolved.

In one respect, however, the iodine pentoxide is unique: it leaves no solid reaction product that may progressively impede the effectiveness of the reactor bed.

Detection in Terms of Acidic Vapors

The sensor responds to the vapors of hydrogen halide, SiF_4, $ClSO_3H$, $RCOCl$, and $CF_3 \cdot CO_2H$. These vapors may therefore be used for neutralizing ammonia and amines in the gas chromatography effluent, with the sensor serving as continuous "back titrator" of the excess acid. Many nonbasic or insufficiently basic organic nitrogen compounds may be converted to ammonia in a hydrogen stream over a nickel contact—the old ter Meulen method of elemental analysis, recently revived for use in coulometric gas chromatography by Martin [26]. This author uses four-electrode electrolytic-potentiometric coulometry. With the galvanic sensor, the train to be recommended is

Hydrogen ———⌐—Column——Ni——⌐—Sensor for acid——→
               ⌐   440°C   ⌐
           Sample   Acid vapor

Organic halogen compounds may be pyrolyzed to yield hydrogen halide. The carrier gas need not necessarily be hydrogen:

Carrier gas————┬———Column———Pyrolyzer———Sensor for acid ——➤
                         │
              Sample

In some cases molecular halogen rather than hydrogen halide will emerge from the pyrolyzer, but this makes no difference to the sensor, which responds identically to, say, HBr and $\frac{1}{2}Br_2$.

Water and compounds with "active" hydrogen should yield equivalent amounts of HCl gas when reacted with a nonvolatile, preferably liquid, readily hydrolyzable organic chlorine compound. The author had some promising results in determining traces of water vapor in a gas stream bubbling through diphenylphosphinous chloride $(C_6H_5)_2PCl$ at 100°C. This liquid reagent was conditioned by bubbling dry air through it for several days at 260°C. Internal condensation reactions produced an oil. Without pretreatment, condensation reactions went on at 100°C, giving a high HCl blank. With the oil, about 75% of the water (37 ppm in 100 ml air/min) was converted to HCl. Although the background was still substantial, the noise level was quite satisfactory (0.15 ppm), but the time for rise and recovery (some 10 min) was not. For purposes of elemental analysis, Belcher et al. [5] and Russian workers [25] have screened hydrolyzable compounds for their ability to convert water (from combusted H) to HCl. The former recommend $\beta\beta'$-diethylglutaroyl chloride, the latter phenylborondichloride. In either case, a cold trap is necessary to separate reagent vapor from the HCl evolved. The ultimate answer to this problem is probably a polymeric material with hydrolyzable Cl.

Several metallic "perfluorides" are known that substitute fluorine for carbon-bound hydrogen, forming HF. The most popular fluorine donor is CoF_3. Passing eluents through a bed of one of these salts at an appropriate temperature should produce transients of HF that may be conveyed to the sensor through Teflon ducts, or even through glass ducts. The SiF_4 formed from the glass would hydrolyze in the cell electrolyte and produce the same signal as HF itself.

Detection with the Sensor for Reducents

The cell is an elaboration of the halogen sensor and determines the gaseous reducents that react readily with bromine or iodine. It is obvious that the cell has potential in the detection of most of those reducents, familiar in classical bromometry and iodimetry, that are volatile. The sensor should be applicable to the selective detection of sulfur-bearing organics subjected to oxidative pyrolysis, yielding SO_2, or subjected to reductive pyrolysis, yielding H_2S.

Use of Two Detectors Simultaneously

In many instances it will be advantageous to split the effluent stream to feed two detector systems simultaneously. The two flow restrictions necessary to ensure a constant ratio of flow rates in the two branches should preferably be located at the exit of the sensors. Both or only one detector system may be of the galvanic family; both or only one system may be of a kind selective for one type of compound.

SENSOR FOR OXYGEN [22]

The galvanic-coulometric sensor for traces of oxygen in other gases is a sandwich structure consisting of three layers: (1) the cathode, borrowed from fuel cell technology, a sheet of porous, inert, conductive material, suitably silver; (2) a diaphragm, in the form of a sheet or sheets of a porous plastic material that immobilizes a restricted volume of aqueous potassium hydroxide solution; and (3) the anode, borrowed from battery technology, a sheet of porous cadmium supported by nickel screen. The cell may be called the hybrid of a fuel cell and a storage battery.

The coulometric performance of this cell, that is its ability to electroabsorb all oxygen offered to it, up to relatively large traces and high flow rates, depends, apart from dimensioning, upon the establishment of micromenisci inside the pores of the cathode. Each meniscus, together with the unflooded portion of the pore wall, is the scene of cathodic dissolution of oxygen. In the older, noncoulometric cells with metal sheet or screen cathodes [13] the shape and extent of meniscus was far less conducive to an efficient uptake of oxygen.

Cadmium is an ideal choice as an anode material because of its position in the electromotive series for aqueous alkali. It is a metal base enough to provide a sufficient electromotive force for the process intended, yet it is not base enough for the reduction of water. An anode metal as base as zinc is unsuitable because it tends to reduce water to hydrogen, thereby generating a parasitic current.

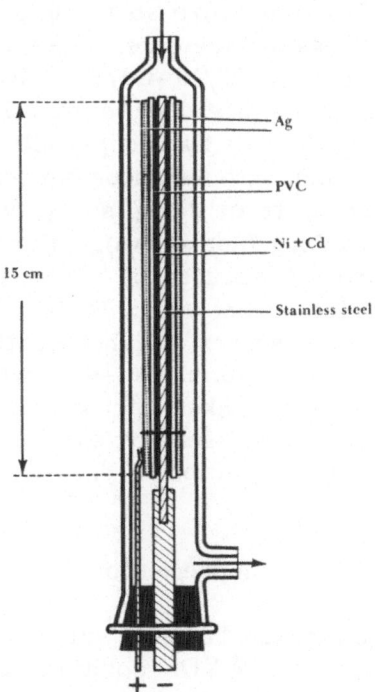

Fig. 2. Sensor for O_2.

Assembly

Numerous cell configurations are possible. Figure 2 shows a double sandwich configuration in a glass housing. A strip of stainless steel (15.0 by 1.5 by 0.1 cm) is inserted between two anode strips (15.0 by 1.5 by about 0.1 cm) cut from Jungner-type alkaline battery negative. A 1.5-cm length of steel strip protrudes at one end. The negative material may be taken from large-size commercial alkaline batteries or purchased as a component. The structure and methods of manufacture of the nickel—cadmium composite have been described [24]. The outer faces of the two anode strips are lined with two strips of porous polyvinylchloride (Porvic S, from Porous Plastics, Dagenham, Essex, England; cut to 15.0 by 1.5 by 0.15 cm). Finally, two strips of porous silver (porosity about 50%, pore size about 30μ; 15.0 by 1.5 by 0.15 cm) are applied against the outer faces of the diaphragms. The seven layers Ag/PVC/Ni—Cd/Steel/Ni—Cd/PVC/Ag are tied together with Nylon thread. The two silver cathode strips are connected to each other by a loop of silver wire. The wire must not make contact with the anode portion of the assembly. It can be prevented from doing so by two short sleeves of capillary Teflon tubing (not shown). The protruding part of the steel "backbone," secured by screw and nut, rests in a slot in the stainless steel rod held by a Neoprene stopper. The outlet rod tube serves as the negative terminal. The positive terminal is a stainless steel wire driven through the stopper. The wire makes spring contact with one of the silver surfaces. The assembly should be tested with an ohmmeter before impregnation to ensure absence of internal shorts.

Impregnation

After assembling, the housing with the element (exit end up) is filled with 24 wt% KOH solution, leaving some head-space. The filter-pump vacuum is applied for a few minutes to remove air from the pores; atmospheric pressure is then readmitted to drive the electrolyte into the pores. After this stage, "charging" of the anode is advisable to ensure that enough cadmium is in the metallic form. To this end, the element is transferred to another tubular container of metal or of glass lined with metal foil or a metal helix. This metal part (stainless steel or nickel) is to serve as the counter electrode during charging. The silver sheets of the element must remain unconnected. The electrolyte is again 24% KOH. Direct current of about 20 mA is applied overnight (16 hr). After charging, the element is withdrawn from the charging vessel and excess electrolyte is allowed to drip off. Further electrolyte must be removed by repeatedly pressing the sandwich between filter paper until no further liquid passes into the paper.

The element is now returned to its housing and tested for continuity by connecting it to a milliammeter. This, with air in the housing, should read 50 mA or more. The air is now displaced by nitrogen, argon, or hydrogen. At first the current declines rapidly. However, a drift-free low background i_0 of a few microamperes is reached only after several hours or overnight. Any persistently high background is almost invariably traceable to in-leakage of oxygen from air (if i_0 is flow-independent) or to oxygen as an impurity in the gas stream (if i_0 is flow-dependent). For most purposes, Tygon sleeves over glass-to-glass junctions in the gas train are permissible. Tygon is not impervious to atmospheric oxygen and contributes to the background, but it does not create signal noise.

Circuit

The cell may be connected to a galvanometric recorder or to a resistor R in parallel with a potentiometric recorder. In the interest of fast response, the load R should be kept as small as possible, as has been explained in a paper on the earlier cell [13]. The load should not exceed 100 ohms. A load of only 1 ohm, giving a time constant of a few seconds, may be used if the potential drop is amplified. Inexpensive solid state amplifiers, suitable for this purpose, have become available lately.

Checks

The coulometric efficiency may be checked by introducing oxygen from an electrolytic source, for example, from the system: Pt wire (anode)/H_2SO_4, 5 N/Hg_2SO_4/Hg, in a configuration shown elsewhere [17, 19]. A constant electrolytic current I evolves oxygen at the platinum wire. The oxygen, swept into the sensor, raises there the galvanic output from a background level i_0 to a level i. With a carrier gas flow rate of $F = 25$ ml/min or less, the coulometric yield $y = (i-i_0)/I$ should be unity. Good cells give complete conversion up to $F = 50$ ml/min or more, at inputs up to $I = 1$ mA or more; 1 ppm O_2 in 25 ml/min generates 6.7 μA.

Another check for coulometric behavior is to use an incompletely purified gas stream, carrying between 10 and 100 ppm O_2, and inserting two cells, A and B, in series. Cell A should show a current increment over its nonflow background; B should show no increment. Upon reversing the flow, B should show the same increment as A did before and A should show none.

Still another check, also with an incompletely purified gas stream, is to vary the flow rate. The increment $i(F) - i(F = 0)$ should be proportional to F.

From Faraday's law, the relation between current increment (μA), oxygen concentration X (ppm by volume, "vpm"), and carrier gas flow rate F (ml/min) is

$$i(X) - i(X = 0) = 0.268 \cdot X \cdot F \cdot \left(1 - \frac{\vartheta - 20}{293} - \frac{760 - P}{760}\right)$$

where F is measured at $\vartheta°C$ and P torr.

Avoiding Dry-Out

In continuous operation of the cell as a monitor, the sample gas stream should carry moisture to prevent the cell electrolyte from drying out. (The first symptom of a "thirsty" cell is sluggish response to oxygen.) A relative humidity of 73%, corresponding to the water vapor pressure over 5 N KOH, is the ideal, but wide deviations are permissible. A convenient method to moisten the gas stream is to pass it along a vertical paper wick that reaches into a small pool of water.

For intermittent work in gas chromatography, such humidification of the carrier gas is not necessary. A shallow pool of cell electrolyte above the bottom stopper humidifies the cell atmosphere during idle periods, when no gas flows, and thereby prevents the element from drying out; or rather, drying out is very much retarded. Very occasionally some make-up water should be added to the pool. A stream of 50 ml/min of dry gas picks up water at a rate of about 0.05 gram/hr. The make-up water may be introduced through the side tubulus while nitrogen sweeps the cell, protecting the element from air. Exposure to air is a shock from which the element recovers only slowly, over a period of hours.

Attenuation

The galvanic increment of the oxygen sensor obeys Faraday's law, that is, the increment is proportional to the oxygen flux $X \cdot F$, only up to certain limits of X and F, limits that depend on the dimensions and geometry of the cell, the microgeometry of the cathode, and the quality of the cathode material. With the dimensions shown and $X \leqq 50$ vpm, $X \leqq 50$ ml/min, coulometric behavior is almost invariably attainable. Sometimes the limits are far wider. However, the oxygen transients may require a base line level above the limit for coulometric performance at the flow rate used. In this case only an aliquot of the oxygen should reach the cell. To meet such a need for an attenuation of oxygen delivery, the gas stream may be split near the entrance of the cell into a minor stream aF and a major stream $(1 - a)F$, and the latter vented, or the major stream may be made to bypass the cell:

$$(1 - a)F$$

Carrier ——⌄—— Column ——⌄—— Combuster ⟍___ aF ___ Sensor for O_2 ——→

Sample O_2

If the attenuation has to be severe, the oxygen in the slow stream aF will reach the cathode only after a noticeable delay, with possible loss of resolution. This can be remedied by feeding the minor stream to a point facing the cathode in the middle of the sensor by means of a narrow capillary that minimizes the volume to be swept through by aF.

Another method of attenuation, one that does not require any modification of the cell design, is to split the stream as before but to route the major portion $(1 - a) F$ through an oxygen-scavenging bed, while the minor portion aF bypasses the bed. The two streams are reunited before reaching the cell. In this way only the fraction a of the oxygen enters the cell, yet with virtually no delay:

$$
\text{Carrier} \longrightarrow \text{Column} \longrightarrow \text{Combuster} \underbrace{\overbrace{\qquad\qquad}^{aF}\atop\underset{(1-a)F}{\text{Scavenger}}} \text{Sensor for } O_2 \longrightarrow
$$

Carrier——Column——Combuster— aF —Sensor for O_2 →
Sample O_2 $(1-a)F$ —Scavenger

In either method, the ratio a is best determined by injecting oxygen electrolytically and comparing the galvanic output $i - i_0$ with the electrolytic input current I: $a = (i - i_0)/I$ Relatively large a's can also be determined by direct flow measurement of aF and F with a soap-bubble meter. This method becomes inaccurate when the attenuation is drastic and a is small.

An excellent material for scavenging oxygen is obtained from a column of granulated electrolytic manganese dioxide (20 to 30 mesh) in a pyrex tube. When reduced *in situ* at 350–400°C in a stream of hydrogen, it turns into pale-green manganous oxide. This product removes from the carrier gas all detectable traces of oxygen at room temperature. For purposes of gas chromatography a column 15 cm high and 1.5 cm in diameter is ample. The material is self-indicating, changing from green to brown when used up. The manganous oxide can be regenerated by repeating the reduction. Instead of hydrogen, a stream of nitrogen loaded with methanol vapor may be employed. After reduction, the bed should be allowed to cool down in a stream of methanol-free nitrogen.

The scavenger bed may also be used to "zero" the cell, by routing all gas through it ($a = 0$, $i = i_0$). A three-way stopcock may provide the alternatives: zeroing, attenuation, and no attenuation, with $a = 0$, < 1, and 1, respectively.

HALOGEN AND RELATED SENSORS

These sensors comprise the system:

Platinum or graphite (cathode)/neutral solution/active carbon (anode)

For halogen itself, the cell electrolyte should be buffered. A typical composition is

KCl............	3.0 moles per liter
KH$_2$PO$_4$	0.1 moles per liter
Na$_2$HPO$_4$.	0.1 moles per liter

Chlorine in the gas stream reacts at the cathode:

$$Cl_2 + 2e \cdot \longrightarrow 2Cl^-$$

Bromine and iodine react similarly. The anodic process takes place on the surface of the carbon and may be written:

$$\ldots C + H_2O - 2e \longrightarrow \ldots CO + 2H^+$$

(The dots signify surface atoms.) Presumably, the carbon surface acquires quinone-like groupings ... CO, and the details are complex. Because of this, the potential at which the anodic event, or events, occur is not as constant and not as well defined as the potential of the cadmium anode in the sensor for oxygen. However, the galvanic current in response to halogen is well defined, depending only on the rate at which the halogen is supplied to the cell. Thanks to the particular choice of anode and electrolyte the cell virtually ignores the presence of oxygen even when the gas passed is air. Two configurations of the device shall be described, a "cavity cell" and a "loop cell." The connections between the source of halogen and the sensors should be of glass only. Metals, plastics (including Teflon), and exposed grease cause losses.

Cavity Cell

In this configuration (Fig. 3) the cathode is exposed partly to the gas and partly to stagnant electrolyte, as in the oxygen sensor. Platinum screen or graphite felt (about 3 mm thick), form a bag lining the inside of an elongated, porous thimble, for example, a ceramic filter candle. The thimble is partly bathed by electrolyte so that it acts as a wick. The anode is a paste of activated carbon and electrolyte. A spiral of platinum screen or graphite cloth, embedded in the sump of paste, collects the current. A capillary glass tube delivers the gas stream at a point in the middle of the cavity. The gas exits through a narrow annular passage between the delivers tube and the cathode material near the mouth of the thimble. The cavity configuration ensures that the halogen impinges directly on the cathode without any opportunity of being adsorbed or absorbed elsewhere—a point especially important in the case of iodine. Water evaporated from the cell should be occasionally replaced.

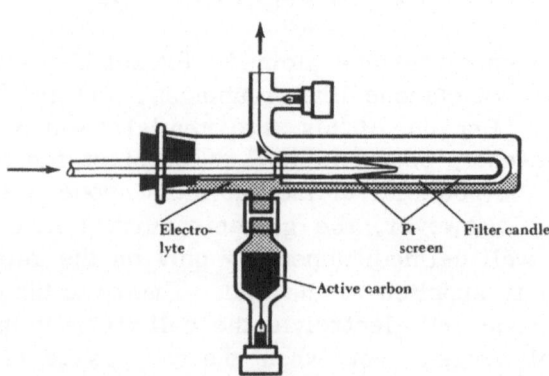

Fig. 3. Cavity cell.

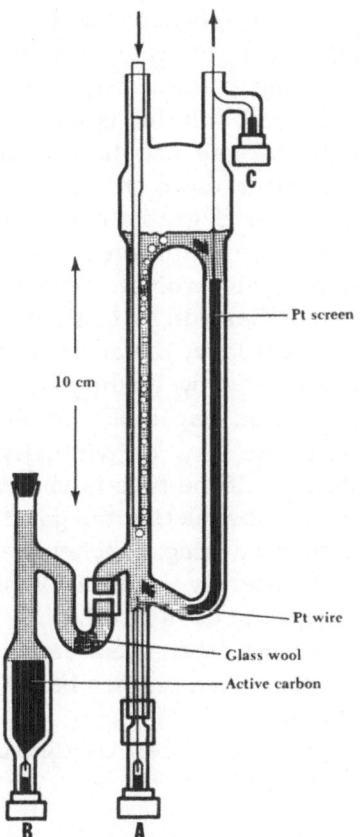

Fig. 4. Loop cell ($A = 3^d$ electrode for reducents).

Loop Cell

In this configuration (Fig. 4) the cathode is completely surrounded by moving electrolyte. The cathode is accommodated in one "leg" of the apparatus, a vertical tube of about 6 mm internal diameter. A platinum screen (6 cm by 13 cm; 20 mesh/cm) is formed into a scroll to fit the leg, or a "tassel" is made of graphite threads. The anode is prepared as for the cavity cell and located in an appendix of the cell, under

stagnant electrolyte.   A second leg of the cell carries a
capillary insert tube through which the gas stream enters.
At the lower end of the insert, the gas emerges in bubbles
that entrain short slugs of solution upward into a bulb-shaped
compartment.   There the gas and liquid separate, the gas to
escape, the liquid to filter down through the platinum scroll
or along and in between the graphite threads.   The inside of
the gas inlet tube should be wetproofed, otherwise halogen can
be retained by adhering electrolyte.   For wetproofing, a film
of a 10% solution of Union Carbide L 31 Silicone Fluid in acetone
is applied to the inlet capillary, the acetone is evaporated, and
the silicone is polymerized by holding the tube at 180°C for
1 hr.   Water repellency is especially important at the tip of
the inlet tube where the dry or virtually dry gas stream
impinges on the solution.   If the tube is untreated, a salt crust
builds up at this point, obstructing the gas flow and retaining
bromine or iodine, thus causing a delayed or even defective
response.   Losses of water by evaporation should be made up
from time to time, as in the cavity cell.

Figure 4 shows a third electrode A serving the analysis of
reducents, as will be described further below, with no function
in the other applications.

The uses to which the two cell configurations may be put
are summarized in Table I.

Table I. Applications of Cavity and Loop Cells

Analyte	Cl_2, t-BuOCl	Br_2, I_2	O_3	Acid vapors	SO_2 and other reducents
Electrolyte	+ buffer	+buffer	KBr + trace KI + buffer	+ KIO_3; no buffer	KBr + trace KI + buffer
Cavity cell	OK	Slightly preferable over loop cell	Not applicable	OK	Not applicable
Loop cell	OK		OK	OK	With third electrode

Sensor for Ozone

The galvanic-coulometric monitoring of traces of ozone in air has been described in an earlier paper, using a slightly different type of loop cell [15]. The electrolyte composition recommended is:

Br^- . . . . 3.0 moles per liter
I^-. . . . . 0.001 moles per liter
$H_2PO_4^-$ . . . . 0.1 moles per liter
HPO_4^{--} . . . . 0.1 moles per liter

The cell is but a sensor for bromine, or rather, tribromide, formed internally through the reaction

$$O_3 + 3Br^- + H_2O \longrightarrow O_2 + Br_3^- + 2OH^-$$

The iodide acts as a catalyst. High concentrations of iodide cannot be used because the oxygen (accompanying the ozone from almost any source) would liberate some iodine and thus cause a blank current.

The cavity cell is not suitable for ozone. Part of the ozone would be decomposed rather than reduced. The ozone must encounter electrolyte only. Metallic parts must be avoided in the connections. All parts exposed to ozone during analysis should be conditioned by several hours' pre-exposure to this gas.

Sensor for Acidic Vapors

For detection of acid vapors, or of other species in terms of an acid vapor, all that has to be changed in the halogen sensors is the electrolyte, which must be unbuffered and must contain iodate. A suitable composition is

Cl^-. . . . . 2.0 moles per liter
I^-. . . . . 0.05 moles per liter
IO_3^- . . . . 0.1 moles per liter

The overall cathodic reaction in response to acids is

$$6H^+ + IO_3^- + 6e \longrightarrow I^- + 3H_2O$$

The anodic reaction in the surface of the activated carbon is the same as with halogen except that the hydrogen ions formed at the carbon surface are not buffered away but are adsorbed, probably in exchange against other cations.

Because iodate is consumed, the electrolyte should be replaced after prolonged use.

Again there is a source-to-sensor transport problem: Any substantial film of moisture on the wall of glass ducts would retain HCl and other acidic vapors as they arrive and release them when they no longer arrive, causing creeping sensor response and recovery. Therefore, before operating, one should flame the glass ducts mildly in a dry stream of air or the ducts should be kept warm with heating tape. The inlet tubes into the cavity and loop cell, of course, cannot be heated, but the capillary bore offers too little adsorptive surface to cause serious delays.

Sensor for Compounds that Reduce Bromine

This is a loop cell (Figs. 4 and 5) with a third electrode A, a short piece of platinum wire serving as the electrolytic anode. The galvanic cathode C is the same as for halogen sensing and has the same function. The activated carbon electrode B in its recess under nonmoving electrolyte acts as a bipolar electrode, namely, as cathode in an auxiliary, electrolytic (externally powered) circuit AB and as anode in the galvanic (self-powered) circuit BC. The electrolyte is a buffered solution of bromide with a trace of iodide, for example,

$$
\begin{array}{lll}
Br^- & 3.0 & \text{moles per liter} \\
I^- & 0.005 & \text{moles per liter} \\
H_2PO_4^- & 0.1 & \text{moles per liter} \\
HPO_4^{2-} & 0.1 & \text{moles per liter}
\end{array}
$$

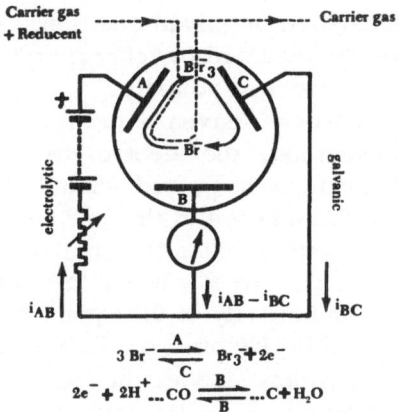

Fig. 5. Principle of reducent detection.

Under an applied emf forcing a constant current i_{AB} through circuit AB, the anode A generates "halogen" (tribromide or triiodide). In the absence of a reducent in the gas stream, all halogen is carried to cathode C and reduced back to halide while filtering through C, thereby generating a galvanic current $i_{BC}(0)$, of the same magnitude as i_{AB}. In the presence of a reducent, part of the halogen is consumed, so that the galvanic current $i_{BC}(X)$ lags behind the electrolytic current. Faraday's law relates the difference to the rate of arrival of the reducent:

$$i_{BC}(0) - i_{BC}(X) = 0.0669 \ n \cdot X \cdot F \cdot \left(1 - \frac{\vartheta - 20}{293} - \frac{760 - P}{760} \right)$$

where X (vpm) is the concentration of the reducent in the carrier gas; F (ml/min) is the carrier gas flow rate measured at ϑ °C and P torr; and n is the number of equivalents per mole reducent.

The difference current may be measured directly in that arm of the "bridge" that leads to the bipolar electrode B.

Without a reducent, the halide concentration and the bipolar electrode suffer no change. In the presence of a reducent, the halide concentration remains unchanged but, as a net effect, the active carbon loses some of its native surface oxygen. It can be restored by adding halogen to the electrolyte and allow-ing this to be consumed in galvanic action. The bipolar electrode may also be rejuvenated by "charging," that is by operating circuit AB with reversed polarity.

Because the reducent leaves a product in the electrolyte, and buffer is consumed, the electrolyte should be renewed occasionally. A running-in period should be allowed during which any reducing impurities in the new solution are given an opportunity to react with anodically supplied halogen.

A monitor for traces of SO_2 in air for ranges down to 0 to 0.5 ppm and with a sensitivity of 0.01 ppm, based on the above principle, has recently become available commercially [4]. Apart from the auxiliary anode, the cell has the same loop configuration as described in the earlier paper on ozone [15], mentioned previously.

By leaving anode A idle, the three-electrode loop cell, of course, may be used for halogen, and with the appropriate electrolyte, for ozone and acid vapors. When changing from other uses to that as a sensor for acid, care must be taken to rinse out the buffer completely.

Attenuation when Using Halogen and Related Sensors

As with the oxygen cell, there are limits of concentration of halogen and of flow rate beyond which the behavior of the halogen sensor is no longer coulometric. In the oxygen sensor, if any oxygen remains "undigested," it simply leaves the cell. Halogen, on the other hand, builds up in the electrolyte, which results in continuous, upward creep of the galvanic signal.

Attenuation by stream splitting may be applied as in the case of the oxygen cell, without or with scavenger. Granulated active carbon is most versatile as a scavenger, but other materials may be chosen, for example, ascarite to retain acid or granulated silver or manganese dioxide to destroy ozone.

The forking point may be located upstream from the reactor or even upstream from the reagent entrance point; the point of reunification may precede or follow the sensor. The following diagrams give some suggestions:

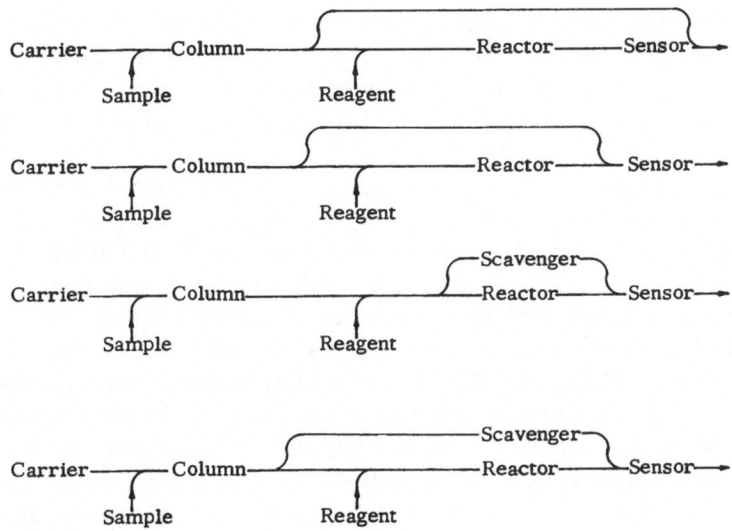

The most trivial way to stay within the range of coulometric performance of any sensor is, of course, to employ a sufficiently small sample or to arrange for a sufficiently slow rate of elution.

REAGENT ADDITION

To benefit from the galvanic–coulometric sensors it is often necessary to introduce reactive gases and vapors into the carrier gas stream at a constant rate. Various "doping" techniques are available: mechanical, evaporative, and (less known) electrolytic. Obvious choices are the motor–driven syringe or bleeds from a lecture bottle. The reactive gas may be added to the carrier directly or to a tributary carrier or to a branched–off portion of the carrier stream, reuniting with it after having picked up the reagent. Only a few methods shall be discussed here.

Electrolysis

For the oxygen decrement method of detecting combustible species, the oxygen required for combustion can be provided by electrolysis or by a deliberate leakage of air. The former requires more apparatus but provides a check on the coulometric performance of the cell.

An electrolyzer for doping gas streams with O_2, based on the system Pt wire/aqueous H_2SO_4/Hg_2SO_4 + graphite/Hg has been described [17, 18]. With reversed polarity, the electrolyzer evolves H_2. The carrier gas with the hydrogen may be led through a drier and then over hot CuO—one means of doping a gas stream with known traces of moisture. Similarly, anhydrous $NiCl_2$ (at 600°C) or $PdCl_2$ (at room temperature after a first heating) converts the H_2 to HCl—one means of providing a base line for detecting ammonia and volatile bases.

The same type of electrolyzer also provides a source for known, constant, and adjustable levels of chlorine. In this case the system to be used is Pt wire (anode)/KCl solution (nearly saturated)/Hg_2Cl_2 + graphite/Hg. The applied current should be kept below about 2 mA, otherwise the migration of the ions would fall behind the consumption of Cl^- at the anode and a portion of the current would generate O_2 instead of Cl_2. Impurities can also depress the yield of Cl_2. The gas line downstream from the chlorine generator should not comprise plastic materials.

The electrolysis of aqueous bromide and iodide produces mainly tribromide and triiodide. For bromine and iodine, an evaporation method should be employed.

Bleed for Oxygen

A simple method of providing a stream of inert gas with a small, constant proportion of oxygen is to route the gas through a tube of hard glass carrying a short length of platinum, a "whisker" fused through the wall. Owing to the mismatch of expansion coefficients, the wire shrinks away from the glass on cooling, opening a minute gap. Atmospheric oxygen diffuses through the gap, driven by the differential of partial pressures, 0.21 atm. This type of diffusion is practically unaffected by fluctuations of ambient temperature, and the level attained is virtually free of ripple. A wire 0.25 mm in diameter fused through a 1-mm-thick borosilicate glass wall admits oxygen from the atmosphere at a rate on the order of 10^{-4} ml O_2/min, corresponding to a galvanic output 27 μA. The magnitude cannot be tailored accurately and occasionally one fails to obtain a gap. A series of "whiskers" in a U-tube gives a choice of several oxygen levels. Any desired number of such a series of bleeds may be closed by raising a column of mercury, outside the U-tube, to the appropriate height. Fine adjustments may be made by slightly throttling the gas flow at a point downstream from the bleed, preferably at the exit from the sensor. This creates a total pressure at the bleed point, or points, in excess of the atmospheric pressure. The result is a minute flow of gas outward that mitigates the diffusion of oxygen inward. Note that the bleed rate is independent of the flow rate of the carrier gas if the pressure of the gas stream at the bleed point is exactly atmospheric. Any flow restrictions downstream not only reduce the bleed rate but also make it flow-dependent.

Another simple method of introducing oxygen for combustion is to route a small, known fraction a of the effluent first through granular activated carbon, then through a short length of thin-walled silicone rubber tubing, and finally to reunite the fractional stream with the main stream before this reaches the combustion chamber. Atmospheric oxygen permeates through the silicone. The silicone would retain sample constituents to an uncontrollable extent, hence the need for a trap. The galvanic decrement must be multiplied by a factor $1/(1-a)$ to correct for the sacrifice of the fraction a of the sample.

Addition by Evaporation

Reagent vapor from liquids like Br_2, t-BuOCl, or $ClSO_3H$ can be bled into the gas line from a bulb bathed in ice and connected to the line via a long capillary bore tube [1]. The rate of ingress of the additive into the carrier gas could be calculated from Gilliland's formula [10] if both vapor pressure and diffusion constant were known. Because sufficient data are not available, one has to resort to trial and error. Trifluoro-acetic acid cannot be handled by this method because of its abnormally low interfacial tension, which causes it to creep up along glass walls.

Iodine vapor can be introduced into the carrier gas by means of a tributary stream of air passing through a column of solid iodine surrounded by ice. An air stream of 10 ml/min (measured at 20°C and 760 torr) entrains iodine at a rate corresponding to a galvanic current of 52.8 μA, as may be calculated from the vapor pressure of iodine, 0.0299 torr at 0°C. Liquid reagents, such as t-BuOCl (estimated $p = 4.3$ torr at 0°C) may be vaporized similarly, for example, by contacting the air stream with pellets of porous alumina im-bibed with the liquid or by passing the air along a wick having one end in a pool of the liquid. The galvanic signal attained is given by

$$i_0 = \frac{nF}{RT}pf = 88.0\ npf \cdot \left(1 - \frac{\vartheta - 20}{293}\right)$$

where F is the Faraday constant; n is the number of equiva-lents per mole reagent; p is its vapor pressure (torr) at 0°C; and f is the flow rate in milliliters per minute of the entraining gas stream, measured at ϑ °C.

Instead of using a tributary gas stream f, a portion aF of the carrier gas F may be split off and led through the saturator. The decrement signal $i_0 - i$ represents, therefore, only the fraction $1 - a$ of the analyte, and $i - i_0$ must be corrected by multiplying with $1/(1-a)$. In the above formula for i_0, the value f for the tributary flow must be replaced by the value aF for the split-off flow.

Hydrogen chloride vapor can be introduced by means of a tributary stream of air bubbled through 64.4 wt% aqueous H_2SO_4 at 25°C (or 61.6 wt% at 20°C) and then through azeo-tropic aqueous HCl (20.221 wt% when distilled at 760 torr), kept in an ice bath.

The HCl pressure of the azeotrope at 0°C is 0.00364 torr. The sulfuric acid bubbler imparts to the air stream the right relative humidity necessary to keep the composition of the pool of hydrochloric acid from shifting.* The galvanic signal in the acid sensor from 10 ml/min air (measured at 20°C and 760 torr) "doped" with HCl in this manner is 32.0 μA. After leaving the hydrochloric saturator, the air stream should be stripped of its moisture in a short column of $CaCl_2$ topped with P_2O_5 and also surrounded by ice. The HCl level can be adjusted over a wide range by adjusting the flow rate of the tributary air stream.

Bromine, with a vapor pressure of 65.9 torr at 0°C, is too volatile a liquid to lend itself conveniently to evaporation at the low rates required for reaction gas chromatography. A thermostated solution of bromine in paraffin or a fluorocarbon solvent may be employed, with a tributary gas stream bubbling through the solution. The rate of pick-up cannot be calculated, only estimated. In theory, it should decline, but such drift will be unnoticeable in practice unless the bromine concentration is very low and the gas flow is very fast. Branching the gas stream, as suggested for iodine, is not practicable with a bubbler. The hydrostatic head adds too much to the dynamic pressure drop in the stream splitter to make stream splitting reliable in this case. Also, the wick or imbibed support method suggested above for a pure reagent is not workable for a binary liquid.

Addition of Ozone

To add ozone to the gas chromatography effluent, a tributary dry air stream may be passed along the "pen" of a "Penray" ultraviolet lamp [15]. Most of the light must be shielded by means of a Pyrex sheath. A 1-mm-diameter pinhole in the sheath transmits enough ultraviolet light to gen-

---

*The air, humidified in the prescribed way, has p_{H_2O} = 2.41 torr when entering the azeotrope. It leaves it with p_{H_2O} = 2.70 torr. The ratio H_2O:HCl in the exit air is then 0.29/0.0364 = 8.0, which is the same as the molar ratio in the liquid. Hence, the liquid suffers no shift in composition.

erate ozone concentrations on the order of 1 ppm, depending on lamp voltage and air flow rate. The lamp with its shield and the air ducts should be conditioned by generating ozone and passing air prior to use, and especially prior to first use, until the galvanic output is constant.

REFERENCES

1. A. P. Altshuller and I. R. Cohen, "Application of Diffusion Cells to the Production of Known Concentrations of Gaseous Hydrocarbons," Anal. Chem. 32:802, 1960.
2. Anonymous "CO on the Highway," Sci. Am., pp. 52 and 57, May, 1965.
3. R. R. Austin, "An Olefin Recorder for Atmospheric Monitoring Service," Trans. Instr. Soc. Am. 1:211, 1962.
4. Beckman Instruments, "Sulfur Dioxide Analyzer," Model 906, Bulletin 4072, 1966.
5. R. Belcher, L. Ottendorfer, and T. S. West, "Acid Chlorides of Substituted Succinic and Glutaric Acids as Hydrolytic Reagents for the Determination of Water," Talanta 4:166, 1960.
6. A. Berton, "Application des Osmopiles Galvaniques à la Détection et au Dosage de Produits Toxiques Volatils," in: J. A. Gautier, ed., Mises au Point de Chimie Pure et Appliquée et d'Analyse Bromatologique, 10e Série, Masson et Cie, Paris, 1962, pp. 6–22.
7. A. Berton, "Gas Chromatography and Selective Galvanic Detection," Chim. Anal. 45:585, 1963.
8. H. P. Burchfield, D. E. Johnson, J. W. Rhoades, and R. J. Wheeler, "Selective Detection Systems in the Analysis of Drugs and Pesticides," in: Lectures on Gas Chromatography-1964, ed. Mattick and Szymanski, Plenum Press, New York, 1964, pp. 129–146.
9. K. Cammann, "Analytical Fuel Cells," Z. Anal. Chemie 216:294, 1965.
10. E. R. Gilliland, "Diffusion Coefficients in Gaseous Systems," Ind. Eng. Chem. 26:681, 1934.
11. M. Guillot and A. Berton, "Analysis of Chlorinated Solvents by Gas Chromatography at Low Temperature with Electrochemical Detection," Comp. Rend. 250:1857, 1960; Chem. Abstr. 55:12156f, 1961.
12. A. J. Haagen-Smit and T. W. Latham, "Carbon Monoxide Levels in City Driving," Clean Air Quart. 8:(4)8, 1964.
13. P. Hersch, "Trace Monitoring in Gases Using Galvanic Systems," Anal. Chem. 32:1030, 1960.
14. P. Hersch, Gas Chromatographic Detector, Brit. Pat. 880, 965, filed February, 1959, published October, 1961.
15. P. Hersch and R. Deuringer, "Galvanic Monitoring of Ozone in Air," Anal. Chem. 35:897, 1963.
16. P. Hersch and C. J. Sambucetti, "Galvanic Coulometric Monitoring of Carbon Monoxide," Paper at Pittsburgh Conference on Analytical Chemistry and Applied Spectroscopy, 1963.
17. P. Hersch, C. J. Sambucetti, and R. Deuringer, "Electrolytic Calibration of Gas Monitors, Analysis Instrumentation, 1963," Proceedings of the Ninth National ISA Analysis Instrumentation Symposium, Houston, Texas, Plenum Press, New York, 1963, pp. 65–71; Chim. Anal. 46:31, 1964.

18. P. Hersch and C. J. Sambucetti, "A Galvanic-Coulometric Monitor of Acid Vapors," Paper at Pittsburgh Conference on Analytical Chemistry and Applied Spectroscopy, 1964.
19. P. Hersch, "Galvanic Analysis," in Advan. Anal. Chem. Instr. Vol. 3 (C. N. Reilly, ed.), Wiley, New York, 1964, p 183–249.
20. P. Hersch and R. Deuringer, "Galvanic Monitoring of Moisture in Gases," Paper at Pittsburgh Conference on Analytical Chemistry and Applied Spectroscopy, 1965.
21. P. Hersch and R. Deuringer, "Galvanic-Coulometric Monitoring of Sulfur Dioxide in Air," Paper at 149th National Meeting of the American Chemical Society, Detroit, 1965.
22. P. Hersch, "Method and Means for Oxygen Analysis of Gases," U. S. Pat. 3,223,597, issued December 14, 1965; Brit. Pat. 913,473, filed March, 1960, published 1962; German Pat. 1,191,984.
23. G. E. Hillman and J. Lightwood, "Determination of Small Amounts of Oxygen Using a Hersch Cell as a Gas Chromatography Detector," Anal. Chem. 38:1430, 1966.
24. W. W. Jacobi, "Batteries, Secondary," in: Encyclopedia of Chemical Technology, Vol. 3, ed. Kirk-Othmen, Wiley, New York, 1963, p. 197.
25. V. A. Klimova and G. F. Anisimova. Izv. Akad. Nauk USSR. Otd. Khim. Nauk 9:2088, 1961; Chem. Abstr. 56:8003g, 1962.
26. R. L. Martin, "Fast and Sensitive Method for Determination of Nitrogen," Anal. Chem. 38:1209, 1966.
27. T. R. Phillips, E. G. Johnson, and H. Woodward, "The Use of a Hersch Cell as a Detector in Gas Chromatography," Anal. Chem. 36:450, 1964.
28. M. Picon, "The Extreme Sensitivity of the Determination of Carbon Monoxide, Ethylene, and Acetylene by Iodic Anhydride," Mém. Soc. Chim. France, Sér. 5, p. 370, 1956; Chem. Abstr. 50:7664a.

Analysis of Some Nonprotein Amino Acids by Gas Chromatography*

K. Blau

*Laboratory for the Study of Neurometabolic Disorders
Biochemistry Department, University of North Carolina
Chapel Hill, North Carolina*

INTRODUCTION

The analysis of amino acids by gas chromatography has become a much more intractable problem than was suspected by the many workers who have tackled it since Bayer's first demonstration in 1957 [1] that various suitable volatile derivatives could be made and separated. Most of the attention has, naturally, been directed toward the analysis of the amino acids which commonly occur in proteins, since the ion–exchange methods presently available set limitations in sensitivity and speed of analysis. Even with the most recent automatic equipment, the number of analyses that can be done in a day is limited, and maintaining the apparatus takes up scarce technical manpower. Gas chromatography holds out great promise in overcoming these drawbacks, but the accuracy of the quantitation achieved by ion-exchange chromatography sets a high standard against which to measure gas chromatographic methods, and this has been approached only recently [2].

*This work was supported in part by Grant 12:HS, Project #236 from the Children's Bureau, Department of Health, Education and Welfare.

This concentration on the protein amino acids has been most helpful because much success has been achieved in chemical methods for preparing volatile derivatives from a group of compounds which are very dissimilar in chemical composition because of the varied chemical structures found in the side-chains. On the other hand, attention has been diverted from the study of many other amino acids of interest, which may be referred to as "nonprotein" amino acids. Although scattered references occur to gas chromatography of volatile nonprotein amino acid derivatives, the group as a whole has been rather neglected. Many of the authors who have published studies on the protein amino acids have included data for one or more nonprotein amino acids. At times, this seems to have been prompted mainly by a natural wish to compare the behavior of all the amino acids they had available. However, at other times it has been recognized that in studies with protein amino acids, nonprotein amino acids are of possible use as markers or as standards in quantitative work, as has been found in the ion-exchange chromatography of amino acids.

There is, however, a great deal of general interest in nonprotein amino acids, since many of them, like ornithine, have a biochemical role to play as intermediates in metabolic pathways. Others may turn up as metabolic end-products in certain conditions and give clues to what may be occurring in the organism, such as β-amino isobutyric acid. New nonprotein amino acids are being isolated and identifield all the time, and a recent monograph on the subject [3] lists over 300 naturally occurring nonprotein amino acids from a wide variety of plant and animal sources. Rapid and sensitive methods are an obvious necessity if these compounds are to be analyzed effectively. Gas chromatography may well be of great use here; it may also aid in the discovery of new nonprotein amino acids because it is much quicker to pinpoint the properties of a compound and to follow its isolation and purification by these rapid and sensitive methods than by almost any other technique. A further reason why nonprotein amino acids repay attention is that it soon becomes evident from a glance at the structures of these compounds that a study of their derivatives by gas chromatography will give much new information about the effect of chemical structure on chromatographic behavior, and this is confirmed by the results obtained from gas chromatography of even a limited number of them. The present article does stress this last aspect, but it should be emphasized that in practice, the other considerations are of paramount importance.

THE CHOICE OF DERIVATIVE

To convert the nonvolatile amino acids into a form suitable for gas chromatography it is necessary to make volatile derivatives from them. This has been approached in many ways: The earliest approach (Bayer, Reuther, and Born [4]) was to use the methyl esters, which were pioneered by Fischer, for separating amino acids

$$\begin{array}{c} COOCH_3 \\ | \\ R\text{-}CH \\ | \\ NH_2 \end{array}$$

by fractional distillation. Another approach was to deaminate the amino acids with sodium nitrite in acetic acid [5],

$$\begin{array}{c} COOH \\ | \\ R\text{-}CH \\ | \\ NH_2 \end{array} + HNO_2 \xrightarrow[CH_3COOH]{NaNO_2} \begin{array}{c} COOH \\ | \\ R\text{-}CH \\ | \\ OH \end{array} + N_2 + H_2O$$

since the polarity of the amino groups has been found somewhat troublesome to handle in gas chromatography. However, this still led to derivatives (hydroxy acids) with two functional groupings to protect in order to prevent cyclization and polymerization reactions and, furthermore, was restricted to a-amino acids. Ninhydrin, a common amino acid reagent, can react with amino acids to produce the aldehydes with one less carbon atom [6],

$$\begin{array}{c} COOH \\ | \\ R\text{-}CH \\ | \\ NH_2 \end{array} \xrightarrow[\substack{or \\ Na\,O\,Cl}]{Ninhydrin} R\text{-}CHO$$

and the same derivatives are supposed to be obtained with alkaline hypochlorite, but what exactly happens with alkaline hypochlorite has been called into question [7]. In any case, the reaction is not sufficiently applicable to amino acids in general.

Currently, the approach is to take each functional group and prepare a suitable derivative. While this increases the molecular weight of the derivatives that are finally analyzed, as compared with the parent compounds, and while this might thus tend to obscure the differences between the components in a mixture, in practice the differences in the chemical structures of amino acids are mostly so pronounced that they are well within the capacity of present-day gas chromatographic techniques.

The most popular way of dealing with amino groups and, generally, basic groups containing nitrogen, is to prepare the trifluoroacetyl (TFA) derivatives.

$$
\begin{array}{ccc}
\text{COOR'} & & \text{COOR'} \\
| & & | \\
\text{R-CH} + (CF_3CO)_2O & \longrightarrow & \text{R-CH} \\
| & & | \\
\text{NH}_2 & & \text{NHCOCF}_3
\end{array}
$$

In these, almost none of the volatility of the amino group is sacrificed, while its polar nature is effectively masked by the electronegativity of the blocking group. Such N-TFA derivatives are very stable, but trifluoroacetylation under the conditions used for their preparation also takes place with hydroxyl and sulfydryl groups to give the corresponding esters and thio-esters of trifluoroacetic acid which are not nearly so stable.

The acid groups are nowadays generally esterified. There is a wide choice of alcohols, but except for specific applications such as the resolution of racemic amino acids, where some of the higher optically active alcohols must be used [8-10], the lower aliphatic alcohols, generally, have been chosen. The TFA amino acid methyl esters

$$
\begin{array}{c}
\text{COOCH}_3 \\
| \\
\text{R-CH} \\
| \\
\text{NH-CO} \cdot CF_3
\end{array}
$$

used in the present study were pioneered by Weygand and his school [11] and, of course, yield the most volatile ester derivatives. This is of particular advantage for derivatives of amino acids with more than one carboxyl group because increasing the chain length of the esterifying alcohol has a cumulative effect in reducing the volatility of their derivatives.

THE CHOICE OF STATIONARY PHASES

The methyl esters of trifluoroacetylated amino acid derivatives do exhibit some tailing in gas chromatography on nonpolar stationary liquids, but on most polar liquids they give symmetrical peaks. The choice of stationary phases is of considerable importance where quantitative work is being done because several polar liquids have recently been implicated [12] in the catalytic breakdown of trifluoroacetylated amino acid derivatives during gas chromatography. The derivatives are those where hydroxyl and sulfydryl groups occur in the side-chain of the parent amino acid and, as mentioned earlier, these groups do not form such stable trifluoroacetates. If quantitative results are required, care must be taken that the type of column used will not catalyze the breakdown of this type of derivative under the analytical conditions being used.

Although it is possible to prepare the volatile derivatives by making first the TFA derivatives of the basic groups [13, 14] and then esterifying with diazomethane, it has been found that the yield in the trifluoroacetylation of amino acids is not always quantitative. However, if the methyl ester hydrochlorides are made first, then a high yield of trifluoroacetylated derivative is obtained from them even under mild conditions, without any need to liberate the amino groups from their hydrochloride form. There are some amino acids where more vigorous conditions are necessary for complete trifluoroacetylation [15], and since it is more convenient to use only one set of conditions, these more vigorous conditions are generally used with the added advantage that they are also quicker. They do not appear to cause breakdown or reduced yields even of such amino acids as tryptophan.

PREPARATION OF DERIVATIVES

Up to 5 mg of the amino acid or mixture of amino acids are put into a $\frac{1}{2}$-dram glass vial fitted with screw-cap with a Teflon liner. If they are in solution they are evaporated to dryness on a rotary evaporator before the next step. For such evaporations, any conventional rotary evaporator is used and fitted with an adaptor to which one can attach a short tube via a 19/38 or 24/40 standard taper ground glass joint. The vial is put inside the tube for evaporation. 1 ml of 5N methanolic HCl is then added to the dry residue (this strength corresponds to a solution which is 50% of saturation with HCl at 0°C, and is madé by diluting a freshly prepared solution of saturated methanolic HCl, made at 0°C, with an equal volume of pure dry methanol). The cap is tightly screwed down and the esterifica-tion allowed to proceed in an oven at 60–70° for 1–2 hr. It is a good idea, to inspect the vial after a few minutes in the oven to check that it is tightly sealed. At the end of the reaction the vial is allowed to cool, the cap is carefully unscrewed, and the contents are evaporated in a rotary evaporator, care being necessary at this point to turn on the vacuum gradually to avoid frothing and sudden bursts of evaporation which would cause loss of sample. The dry residue of the methyl ester hydro-chlorides is trifluoroacetylated with 0.1 ml trifluoroacetic anhydride in 0.4 ml of methylene chloride in the same sealed vial for 10 min in an oven at 150° [15]. After cooling, this solution is ready for immediate injection. It is not advisable to evaporate the trifluoroacetylation mixture, since this has been shown to cause loss of the more volatile components of a mixture [16]. The structures of some of these amino acid derivatives have been checked by putting the derivatives through a combined gas-chromatograph–mass-spectrometer. In every case, the fragmentation patterns obtained have been entirely consistent with the structures that one would expect from the above reaction sequence. In most cases, it was also possible to obtain the correct molecular weight from the molecular ion.

In the present work, most of the amino acids were weighed

out to correspond to a total injection of 1–5 μg of the parent amino acid in 1 μl of solution.

INSTRUMENTAL DETAILS

Most of the work described here has been done with dual-column, dual flame-ionization detector instruments (Micro Tek MT 220 & F & M 402) which have performed very well. Some of the earlier work was done on single column instruments of various other types, all of which gave essentially similar results. However, it is not advisable to use argon β-ionization detectors for these derivatives, as the TFA groups are, to some extent, electron capturing, and where an amino acid forms derivatives with more than one TFA group, one may get only very small peaks or even negative deflections below the base line with β-ionization detectors; all of these factors are confusing and make it very difficult to get consistent quantitative results.

It has been found in this work that a dual-column chromatograph can be used essentially as a twin-column chromatograph even with only a single-pen recorder. For this mode of operation, the recorder zero is set at the center of the chart and the electrometer is also balanced to give a base line at chart center. Injection of samples may then be made into either column, but the peaks will be recorded as deflections to left or right of the base line, depending on which column is concerned. In practice, there is almost never a situation where two peaks exactly cancel each other out even if they emerge together on both columns, and the interpretation of the chart record presents no practical difficulty even where the peaks are simultaneous (for an example, see Fig. 12), although in such cases it is not possible to measure the peak heights or areas.

By using the instrument in this way, we also have the advantage in that columns with different characteristics can be used simultaneously. In the present work, for example, some of the separations have been obtained on two columns used in the way just described: one is a nonpolar silicone (4% DC-710) and the other a polar silicone (2% XE-60). Since many of the amino acid derivatives still have some residual polar character, they often exhibit longer retention times or higher retention temper-ature on polar as compared with nonpolar stationary phases, and so, the more polar column is loaded with only 2% stationary phase. The 4% loading on the nonpolar column then gives more comparable results when the two columns are working together, and it also tends to sharpen up the peaks and reduce tailing. These columns are quite well-matched, and even with temper-ature-programming a constant base line is obtained up to about 200°, although above that temperature there is more bleed from the more heavily loaded column.

Most of this work has been concerned with a survey of the chromatographic behavior of a wide range of compounds whose structures were very varied. In many cases, there was an initial question of whether a volatile derivative had, in fact, been prepared. For these reasons, temperature programming was used extensively, since any volatile derivative that has been made, unless it has very low volatility, breaks down during chromatography, or is irreversibly absorbed on the column, has to emerge as a peak at some point of the program.

Further details of the particular columns used and of the operating conditions for any given separation will be found in the captions to the text figures.

RESULTS AND DISCUSSION

Figure 1 shows peaks obtained from a number of a-amino acids with unbranched hydrocarbon side-chains of increasing length. The effect of ascending a homologous series is re-flected in the increasing retention parameters and is quite pro-nounced with a-amino octoic acid. It is clear that glycine, which ought to be the first member of this series, emerges much later than would therefore be expected. This is because it has two hydrogen atoms on the a-carbon atom, and this methylene grouping confers greater polarity on the molecule as a whole. On a nonpolar column such as SE-30, this polar behavior of glycine is not so pronounced, although the glycine derivative is still eluted after the alanine one.

In this discussion, the term "polar" is used rather loosely both as applied to stationary phases and to derivatives that are separated by gas chromatography. What is generally meant is that "nonpolar" columns separate compounds predominantly by means of differences in their volatility, whereas "polar" columns, in addition, exert an additional and in many cases an overriding selectivity according to the chemical nature of the stationary phase and of the compounds that are being separated on them. In this sense, polarity implies the presence of groupings within the molecule such as alcohol, ketone, acid, basic, ester, amide, etc., which on reacting with similar groupings of the stationary phase will cause a retention which

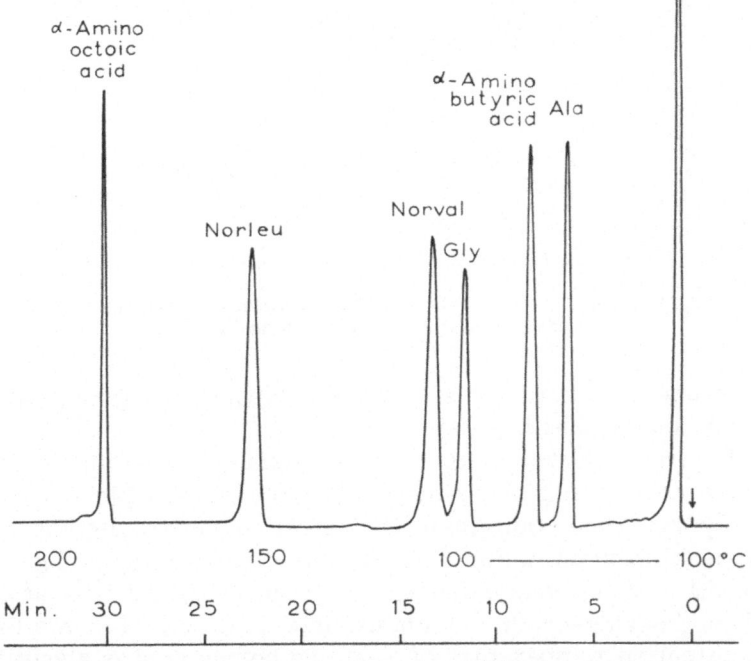

Fig. 1. Separation of straight-chain α-amino acid derivatives. Injection at arrow. Conditions: 2 m × 4 mm glass U-tube column packed with 5% XE-60 on 90–100 mesh Anakrom ABS. Gas flow 38 ml N$_2$/min. Inlet 220°; detector 250°; temperature program: 100° for 12 min, then 5°/min. to 200°.

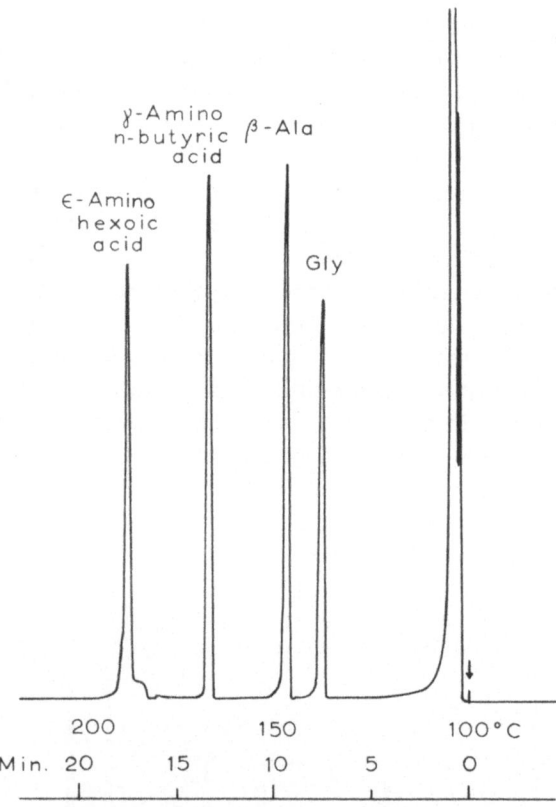

Fig. 2. Separation of ω-amino acid derivatives. Conditions as in Fig. 1. Temperature program: 100° + 5°/min to 200°.

is, of course, characteristic of the compound being run and of the stationary phase.

Figure 2 shows peaks from a number of ω-amino acids, and here one sees superimposed on the effect of lengthening the hydrocarbon side-chain the effect of separating the two functional groups: the methyl esterified carboxylic acid group and the trifluoroacetylated amino group which is now at the end of the side-chain. A comparison with Fig. 1 shows that the retention temperature of ε-amino hexoic acid is about the same as that of the α-amino octoic acid derivative, so that the effect of separating the amino from the carboxyl group is evidently greater than that of lengthening the side-chain. The same effect is brought out even more clearly if we look at the isomeric amino butyric acids in Fig. 3, where the only differences between the derivatives in Fig. 3a (straight chain) or in Fig. 3b (branched chain) are in the distance between the amino and carboxyl derivatives within the molecule.

The effect of separating these groups applies also in the aromatic series, as exemplified by the behavior of the three amino benzoic acid derivatives. Figure 4 shows that they are eluted in the sequence that one would expect. However, the

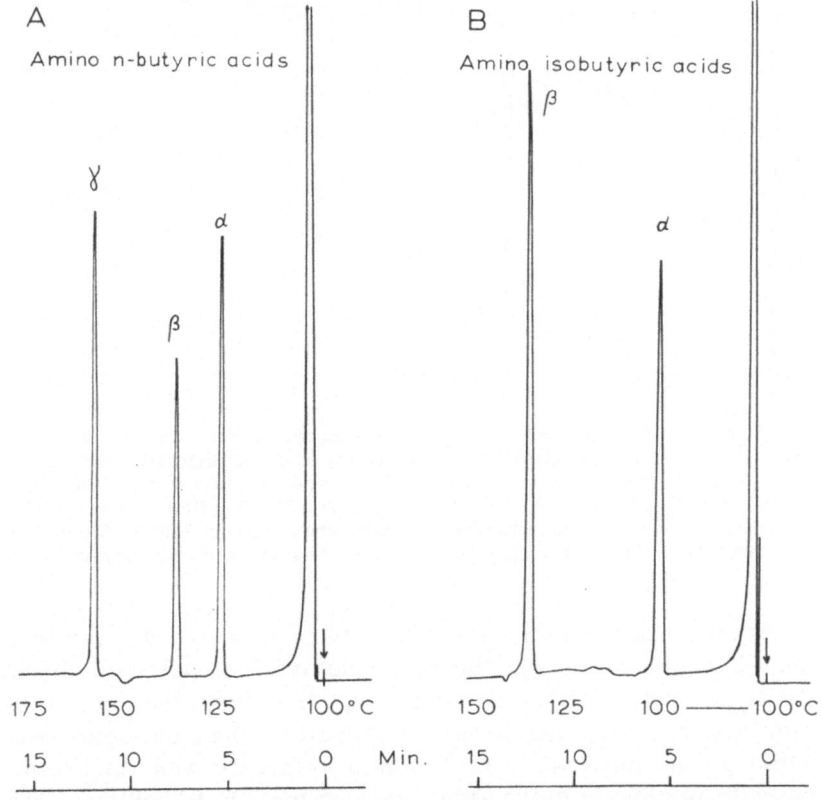

Fig. 3. Separation of the isomeric amino butyric acid derivatives. A. Isomeric amino n-butyric acids. B. Isomeric amino isobutyric acids. Conditions as in Fig. 1. Temperature programs: 100° + 5°/min as indicated.

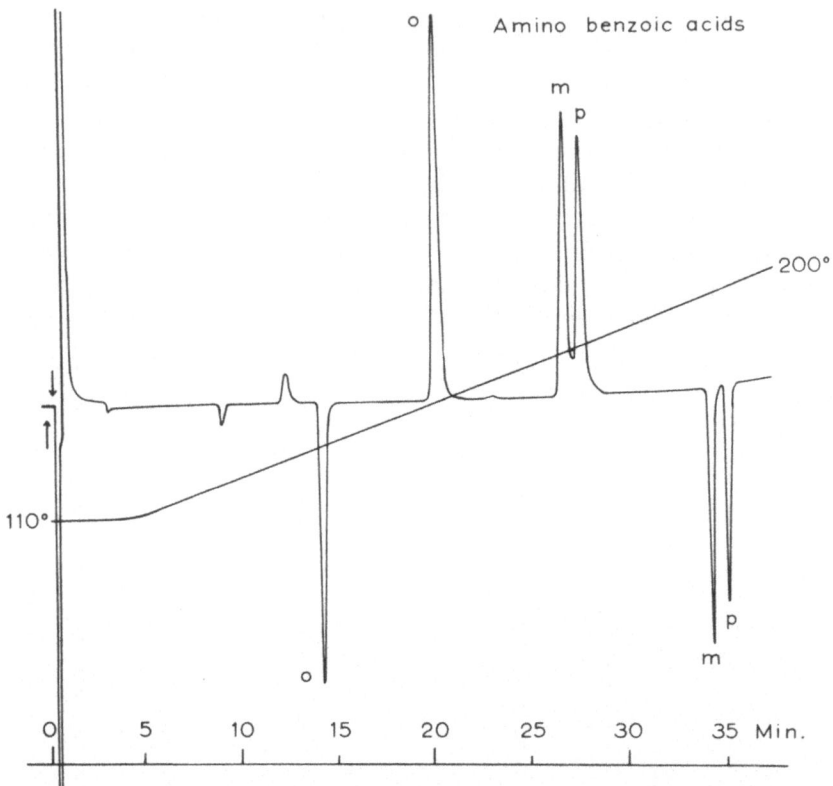

Fig. 4. Separation of isomeric amino benzoic acid derivatives. Conditions: upward
peaks on 3.6 m × 2.5 mm glass columns in the paper-clip configuration packed with
4% DC-710 on 90-100 mesh Anakrom ABS; downward peaks on matched column
packed with 2% XE-60 on 90-100 mesh Anakrom ABS. Gas flow 40 ml He/min in
each column. Inlet 220°; detector 250°; temperature program: 110° for 5 min, then
3°/min to 200°. The temperature was recorded continuously during the separation.

differences between the meta and para isomers are far less
than between them and the ortho derivative, particularly on
the more polar XE-60 column. On this column the ortho isomer appears at a far lower temperature than the other two.
Although in the less polar column volatility, was important,
polarity is clearly more significant on the XE-60 column. The
two substituents in the ortho position of o-amino benzoic acid
probably interact with one another even after derivative forma-

tion, so that they are then not able to interact with the polar groupings of the stationary phase. Such an intramolecular interaction would not be possible in the meta or in the para isomer. This explanation is supported by the relative positions of the ortho derivative; this compound comes off later on the 4% DC-710 than on the 2% XE-60 column, as one would expect only if the compound had little polar character. The meta and para derivatives reverse this behavior, clear indication that these derivatives must be much more polar. In this manner, the simultaneous use of a nonpolar column can be helpful in the interpretation of gas chromatographic results.

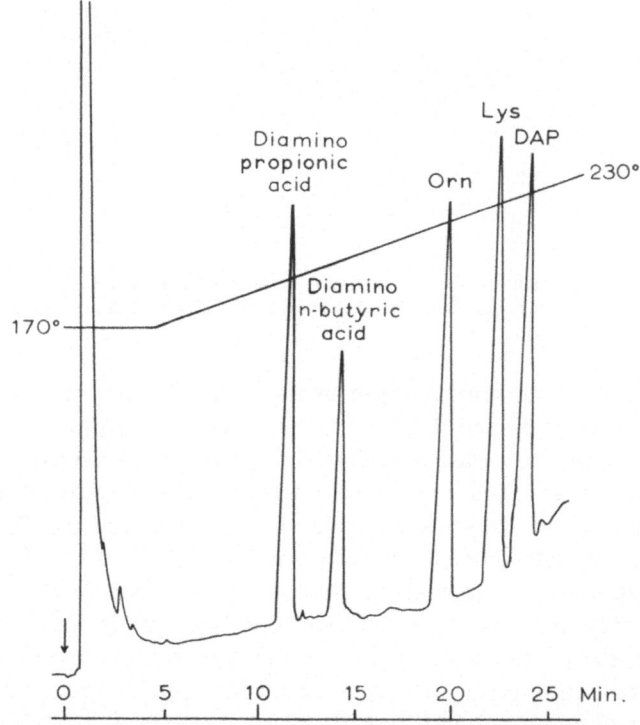

Fig. 5. Separation of the derivatives of some basic amino acids and of diamino-pimelic acid (DAP). Conditions: 3.6 m × 2.5 mm glass column in the paper-clip configuration packed with 2% XE-60 on 90-100 mesh Anakrom ABS. Gas flow 25 ml He/min. Inlet 220°; detector 250°; temperature program: 170° for 5 min, then 3°/min to 230°.

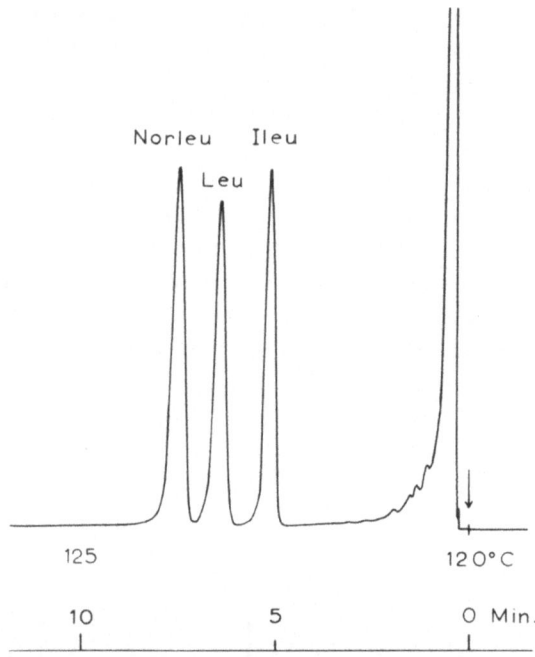

Fig. 6. Separation of the derivatives of the isomeric leucines. Conditions as in Fig. 1. Temperature program: 120 + ½°/min to 125°.

The basic amino acids in Fig. 5 also demonstrate the effect of separation and increased chain length, although here there is a considerable increase in retention behavior due to the extra basic grouping present within the molecule, even though it is trifluoroacetylated. Diaminopimelic acid (DAP) has been included with this basic group, as it is linked to bacterial lysine metabolism. It is useful to be able to separate DAP from lysine not only for that reason, but also because it seems that DAP may turn out to be a useful marker for detecting situations where bacterial action has occurred, since its presence has not been demonstrated other than in bacteria [17]. The peaks of the derivatives show "heading" due to column overloading. They were run at rather a high temperature program, and the signal was strongly attenuated to keep base line drift from going off the chart.   In order to show the peaks, larger than usual samples (about 20 μg) were applied. These derivatives (except DAP) tailed badly on the DC-710 column.

A general rule based on widespread experience, for example, in the gas chromatography of hydrocarbons, is that branching within the molecule increases the volatility of a compound so that it elutes sooner than the corresponding unbranched compound. The same effect is found with branched side-chain amino acids.   This can be seen if one compares the amino n-butyric acid iso-butyric acids in Fig. 3.   It shows up more clearly in Fig. 6, for the isomeric leucines, and in Fig. 7, for the isomeric valines.   The separation of the valines is more widely spaced because there is no intermediate form between norvaline and valine and because isovaline has an α-methyl group instead of an α-hydrogen.

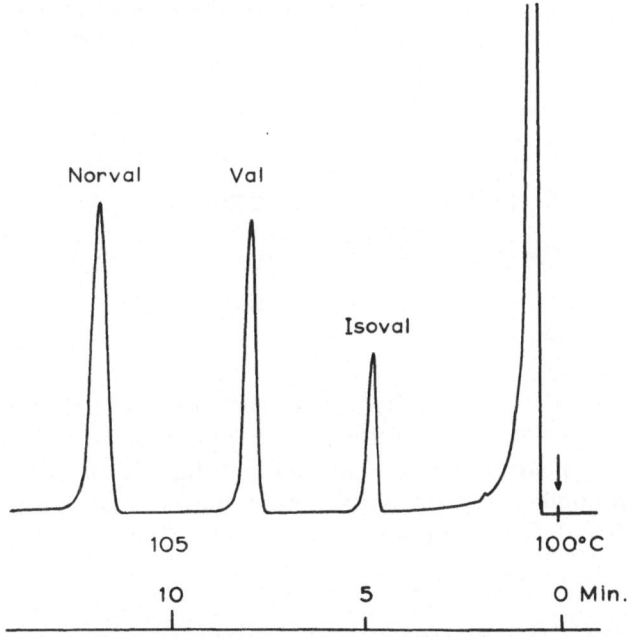

Fig. 7. Separation of the derivatives of the isomeric valines. Conditions as in Fig. 1. Temperature program: 100° + ½°/min.

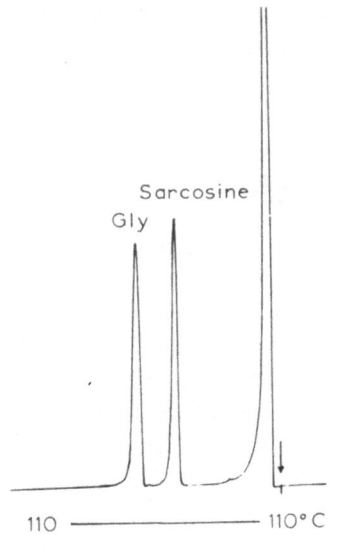

Fig. 8. Separation of the glycine and sarcosine derivatives. Conditions as in Fig. 1. Isothermal at 110°.

$$CH_3-CH_2-CH_2-\underset{\underset{NH_2}{|}}{\overset{\overset{COOH}{|}}{CH}}$$

NORVALINE

$$\underset{CH_3}{\overset{CH_3}{>}}CH-\underset{\underset{NH_2}{|}}{\overset{\overset{COOH}{|}}{CH}}$$

VALINE

$$CH_3-CH_2-\underset{\underset{NH_2}{|}}{\overset{\overset{COOH}{|}}{C}}-CH_3$$

ISOVALINE

The absence of the a-hydrogen reduces the polarity of the derivative, so that it is less strongly retained by the polar stationary phase XE-60. The same thing has been observed with a-methyl glutamic acid, which emerges before glutamic acid, although it has a higher molecular weight.

$$HOOC-CH_2-CH_2-\underset{\underset{NH_2}{|}}{\overset{\overset{COOH}{|}}{CH}}$$

GLUTAMIC ACID

$$HOOC-CH_2-CH_2-\underset{\underset{NH_2}{|}}{\overset{\overset{COOH}{|}}{C}}-CH_3$$

a-METHYL GLUTAMIC ACID

These separations of the leucines and the valines are not possible on nonpolar stationary phases, on which these two sets of isomers are not completely resolved.

A different kind of methyl group substitution is the methylation of the amino group. Figure 8 shows the separation of glycine and N-methyl glycine (sarcosine). N,N–dimethyl glycine methyl ester has also been prepared; this derivative, since it is a tertiary amine, does not need to be trifluoroacetylated. When run under the conditions of Fig. 8 it emerges at 75°, i. e., 35° lower than the N-TFA sarcosine methyl ester.

We have studied the gas chromatography of the N,N–dimethyl amino acid methyl esters derived from a number of amino acids [18] because these are some of the most volatile derivatives of amino acids. There are various sequences for preparing them from the amino acids, the most convenient of which is catalytic hydrogenation in excess formaldehyde solution [19] followed by esterification with diazomethane:

$$
\begin{array}{ccccc}
\underset{\substack{|\\ R-CH\\ |\\ NH_2}}{COOH} & +\ 2\ HCHO & \xrightarrow[Pt]{H_2} & \underset{\substack{|\\ R-CH\\ \diagup N \diagdown\\ CH_3\ CH_3}}{COOH} + CH_2N_2 & \longrightarrow & \underset{\substack{|\\ R-CH\\ \diagup N \diagdown\\ CH_3\ CH_3}}{COOCH_3}
\end{array}
$$

Unfortunately, like many procedures in the amino acid field, this method only works well with the simpler amino acids, and the predominance of side reactions with sulfurcontaining amino acids and with tryptophan ruled out this approach as a general method. Direct methylation reactions could not be used, since these lead to a mixture of products. For any satisfactory chromatographic method "one compound—one peak" is the safest rule. Methylation ultimately leads to the formation of betaines, and these are quaternary ammonium salts, and therefore no longer volatile.

$$
\underset{\substack{|\\ R-CH\\ |\\ NH_2}}{COOH} \xrightarrow{CH_3X} \underset{\substack{|\\ R-CH\\ |\\ NH\\ |\\ CH_3}}{COOH} \longrightarrow \underset{\substack{|\\ R-CH\\ \diagup N \diagdown\\ CH_3\ CH_3}}{COOH} \longrightarrow \underset{\substack{|\\ R-CH\\ |\\ N^+(CH_3)_3}}{COO^-}
$$

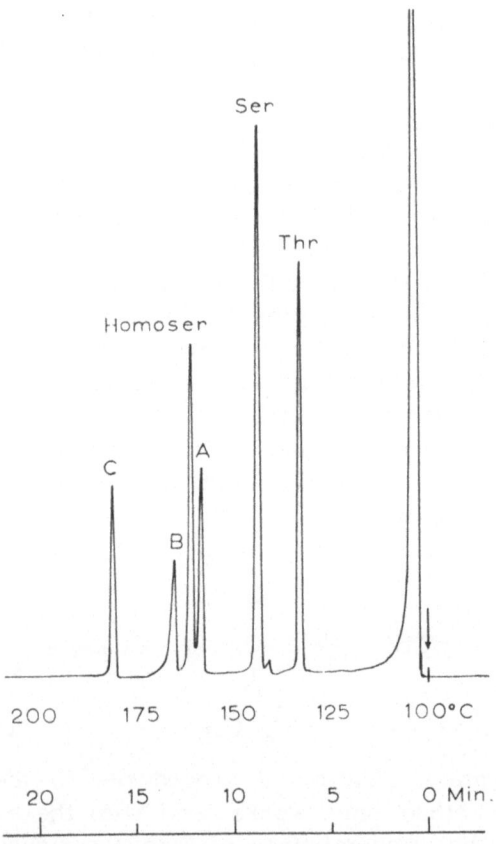

Fig. 9. Separation of the derivatives of some hydroxy amino acids. Conditions as in Fig. 1. Temperature program: 100° + 5°/min to 200°. The peaks labeled A, B, and C are the N-mono TFA derivatives corresponding to the threonine, serine, and homoserine derivatives, respectively, and derived from them by hydrolysis (see text).

Dimethylamino acid esters are tertiary amines and show a tendency to give peaks which tail. Although there are now a number of ways by which tailing of polar solutes can be minimized, it seems preferable to concentrate on derivatives which do not give tailing peaks.

Peaks from some amino acids with hydroxyl groups in their side chains are shown in Fig. 9. The threonine derivative almost invariably precedes the serine derivative, although it is the next higher homologue and might thus be expected to elute later.

$$\begin{array}{ccc}
\text{COOH} & \text{OH COOH} & \text{COOH} \\
| & | & | \\
\text{HO-CH}_2\text{-CH} & \text{CH}_3\text{-CH-CH} & \text{HO-CH}_2\text{-CH}_2\text{-CH} \\
| & | & | \\
\text{NH}_2 & \text{NN}_2 & \text{NH}_2 \\
\text{SERINE} & \text{THREONINE} & \text{HOMOSERINE}
\end{array}$$

There are two possible explanations for this: first, there may be a difference in polarity between the primary and secondary alcohol groups; second, in serine the β-carbon atom is present as a methylene grouping and is more polar than the β-carbon atom of threonine with only one hydrogen. Homoserine, which is actually isomeric with threonine, is even less volatile. The chart shown in Fig. 9 shows three small extra peaks which have arisen from the sample on standing and which were not present originally.   These have been included purposely to illustrate that the derivatives of the hydroxy amino acids are easily hydrolyzed [14, 20, 21].   Only the trifluoroacetylated hydroxyl group is attacked initially, since this is sensitive to moisture. The products, which are still volatile and give rise to the extra peaks in Fig. 9, are the N-mono TFA derivatives with free -OH groups, which always have longer retention parameters.   On prolonged standing, the original di-TFA derivatives disappear almost entirely; on further stand- ing the mono TFA derivatives also decrease, and it is assumed that this is due to the hydrolysis of the equally susceptible O-mono TFA derivatives, which can be formed by an intra- molecular N → O acyl shift of the TFA group [20].

Some sulfur amino acid derivatives are shown in Fig. 10. The cysteine derivative is the N,S–diTFA cysteine methyl ester. The S-TFA group is even more easily hydrolyzed than the O-TFA groups, and N,S–diTFA cysteine can act as a trifluoroacetylating agent [21], rather like TFA ethanethiol which was used by Schallenberg and Calvin [22]. At one time, it was thought that methionine gave the sulfoxide during esterification [23], but this was an error based on the fact that the chroma-

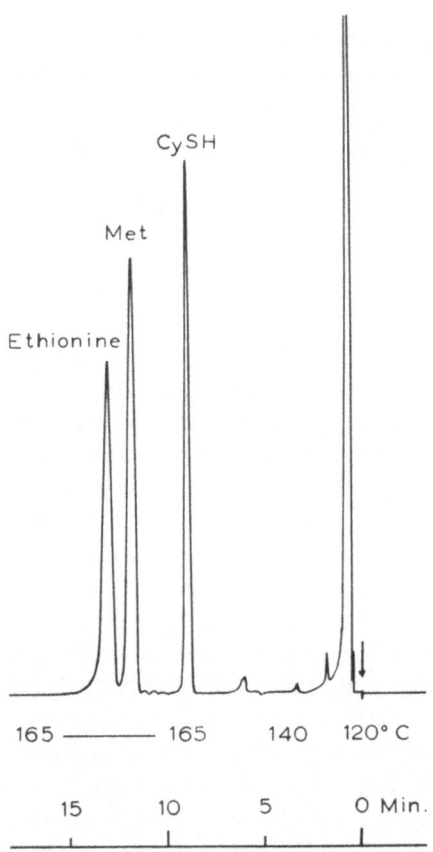

Fig. 10. Separation of the derivatives of some sulfur-containing amino acids. Conditions as in Fig. 1. Temperature program: 120° + 5°/min to 165°, then held at 165°.

tographic properties of the sulfoxide are very similar. The corresponding sulfone has a very much higher retention temperature and is not included. Cystine also comes off at a much higher temperature and, further, usually gives low yields and a broad tailing peak. It is the last and least satisfactory in the gas chromatography of the protein amino acids. A comparison with diaminopimelic acid suggest that the –S–S– bond may be very low in volatility. The thioether linkage also appears to be relatively nonvolatile, and so far it has not been found possible to get peaks from lanthionine, cystathionine, or djenkolic acid.

A number of cyclic amino and imino acids have been investigated, and a separation attempt is shown in Fig. 11. The 4%
DC-710 column is not very good for this because it does not
resolve proline, hydroxyproline, and pipecolic acid. On the 2%
XE-60 there is almost complete resolution, except for proline
and cycloleucine.

PROLINE   CYCLOLEUCINE   PIPECOLIC ACID

It is interesting that the pipecolic acid derivative, with a
larger ring, and as seen with the nonpolar column, with comparable volatility, should come off before the proline derivative
on the polar column. This may be because of the two heterocyclic bases from which these two compounds stem pyrrolidine
is a slightly stronger base than piperidine.

It can be seen that pyroglutamic acid (pyrrolidone carboxylic acid) and glutamic acid are separated on both columns,
although their order of emergence is reversed.

PYROGLUTAMIC ACID   GLUTAMIC ACID

The glutamic acid, from which pyroglutamic acid is usually
prepared, is either an impurity in the pyroglutamic acid or
arises from it during the esterification reaction.

With the separation of the aromatic amino acids in Fig. 12
we see a situation where peaks are eluted simultaneously from

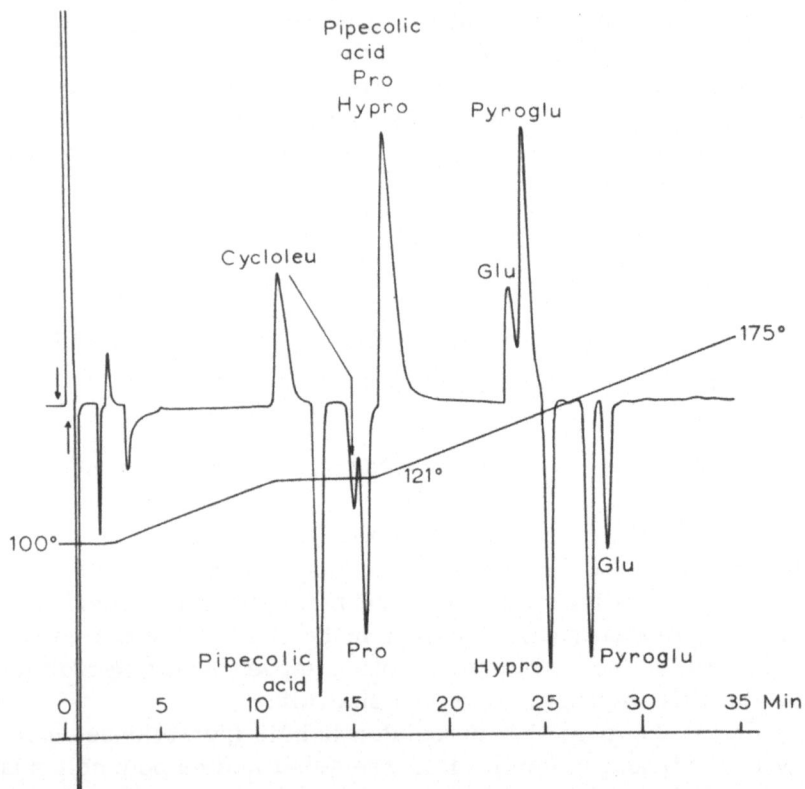

Fig. 11. Separation of some derivatives containing ring systems. Conditions as in Fig. 4. Temperature program: 100° for 2 min, 3°/min to 121°; held at 121° for 5 min then 3°/min to 175°.

both columns, and the patterns obtained with a-phenylglycine and phenylalanine are characteristic of the results one can get. The relative heights of the peaks on either side of the base-line depend, of course, on the relative amounts of compound in each peak. On the polar XE-60 column there is an increase in the retention temperature as the number of hydroxyl groups (here trifluoroacetylated) substituted into the ring increases. However, the difference between phenylalanine and tyrosine is much greater than between tyrosine and 3,4-dihydroxyphenylal-anine (DOPA), presumably because of the type of positional effects discussed earlier under the amino benzoic acids. Generally, it is found that substitution of a trifluoroacetylated hydroxyl group increases retention time or temperature, as is seen in a comparison of alanine and serine, or proline and hydroxyproline (Fig. 11). On the other hand, if one substitutes the rather more polar amino group into tyrosine, one obtains a greater increase in both polarity and boiling point of the derivative. In fact, the 3-aminotyrosine derivative does not come off the XE-60 column at all, not even isothermally after the temperature has leveled off at the top of the temperature program. The derivatives of the compounds which have phenolic

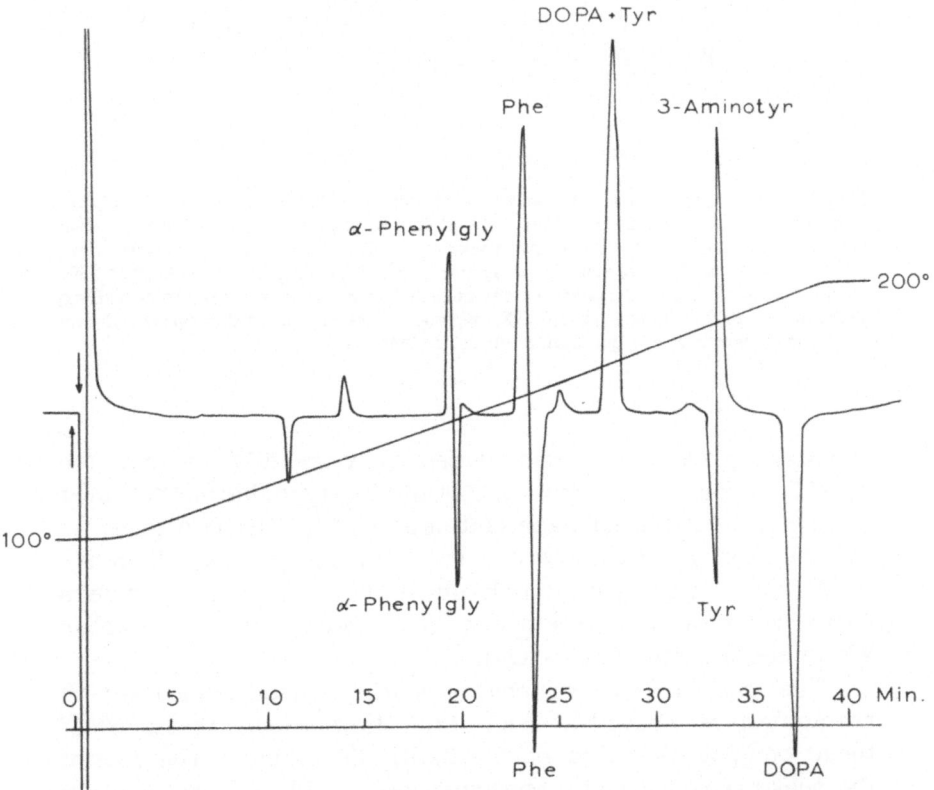

Fig. 12. Separation of derivatives from some aromatic amino acids. Conditions as in Fig. 4. Temperature program: 100° for 3 min then 3°/min to 200°.

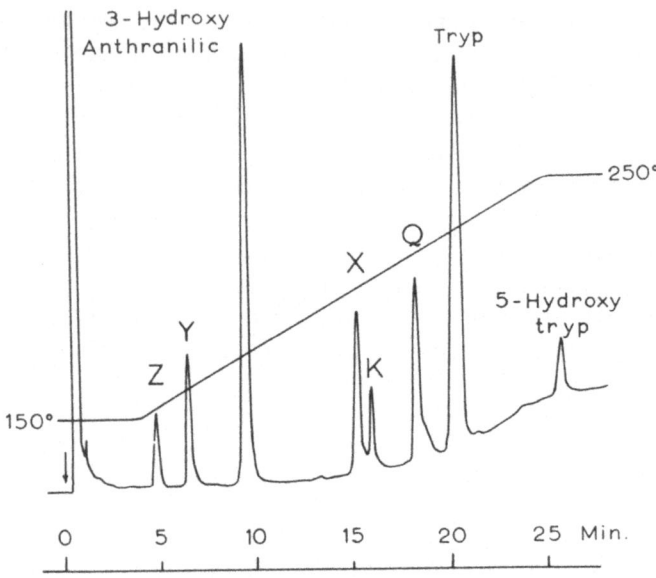

Fig. 13. Separation of the derivatives of some compounds related to tryptophan. Conditions: 3.6 m × 2.5 mm glass column in the paper-clip configuration packed with 4% DC-710 on 90-100 mesh Anakrom ABS. Gas flow 25 ml He/min. Inlet 220°; detector 250°; temperature program: 150° + 5°/min to 250° then held at 250°. Peaks labeled Z and Y are unknown compounds derived from the sample of hydroxy anthranilic acid. Peaks labeled X, K, and Q are due to the derivatives of xanthurenic, kynurenic, and quinaldinic acids, respectively.

hydroxyl groups are hydrolyzed very readily to lose the -O-TFA grouping and, in fact, O,N-diTFA-tyrosine methyl ester is an excellent trifluoroacetylating agent [21]. The TFA group is not the best protecting group for -OH compounds, which are preferably trimethylsilylated, since the trimethylsilyl ethers are much more stable to hydrolysis. This is one of the areas which needs further investigation.

The gas chromatography of some compounds related to tryptophan is shown in Fig. 13. It was found that most of these compounds tailed on the 2% XE-60 column. The size of the peaks from kynurenic and xanthurenic acids is much smaller than it should be.

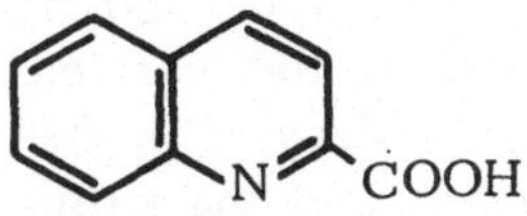

QUINALDINIC ACID·

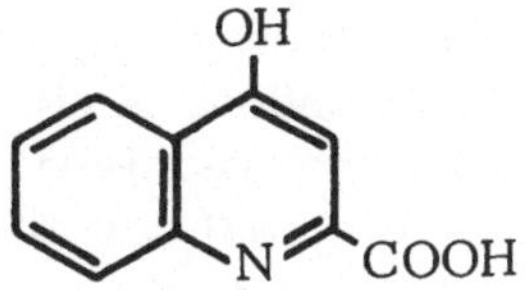

KYNURENIC ACID

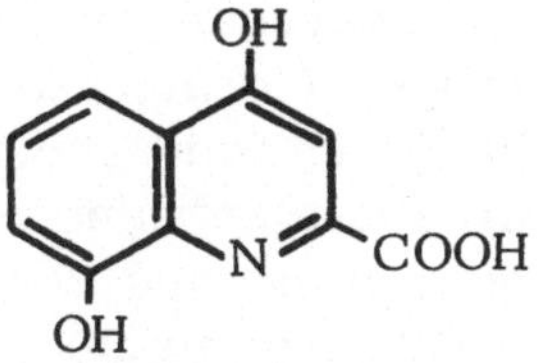

XANTHURENIC ACID

Again, this is probably due to the ease of hydrolysis of these –O-TFA derivatives, as already mentioned in connection with the hydroxy amino and phenolic amino acids. It is surprising that the hydroxylated quinaldinic acids should come off before quinaldinic acid itself. However, this applies only to the non-polar DC-710 column: no peaks were seen for the hydroxylated compounds on the XE-60 column. As suggested above, the trimethylsilyl group is probably the most suitable group for protecting these phenolic hydroxyl groups.

CONCLUSION

Nothing has been said in this report about the quantitative determination of these compounds. It is admitted that their gas chromatographic analysis has not yet been put on a sound quantitative basis. The present preliminary survey has been concerned mainly with seeing how many nonprotein amino acids could readily be converted into volatile derivatives and how easy it would be to separate these derivatives by gas chromatography. It should perhaps be made clear that these derivatives were all run separately and individually under the same conditions as those used for the separations, so that there is no uncertainty about the identification of the peaks. As is apparent from the figures, most of the protein amino acids figure in the separations wherever appropriate; their presence helps to put the analysis of nonprotein amino acids within the framework of amino acid analysis in general.

There are still great gaps in this subject: a very limited range of compounds has been covered. For example, nothing has been said about histidine and the imidazole derivatives, or about arginine and guanidine compounds. Very little has been presented concerning indole derivatives, and the sulfur containing amino acids have received only a brief glance. Work is proceeding on the analysis of the protein and non-protein amino acids of these and other groups by gas chromatography. However, derivative formation is not so straightforward in this area; often side reactions occur, which may degrade the parent compound, depress yields in derivative formation, or lead to multiple products. Some types of amino acids yield derivatives which are still polar, which are not very stable or very volatile, or a combination of some or all of these. In such cases, one never sees a peak and is left to wonder why. Other problems may arise because of limitations inherent in stationary phases or columns or instrumentation.

A great deal of work is still required: research into the use of new derivatives; development work on columns and evaluation of new supports, new stationary phases, and the use of mixed stationary phases. And there is the vital problem of quantitative studies to find ways of improving the yields in derivative formation and suitable methods of measuring the peaks for estimation procedures. The nonprotein amino acids, whose numbers are steadily increasing, offer a stimulating challenge to those who are intent on applying gas chromatographic methods to their analysis.

ACKNOWLEDGMENTS

Some preliminary results of this work were presented at the 4th Symposium on Gas Chromatography, École Polytechnique, Paris 1965.

I should like to thank my former colleague Dr. A. Darbre of King's College, London, most warmly for all his wholehearted support and encouragement.

REFERENCES

1. E. Bayer in: Gas Chromatography 1958, ed. D. H. Desty, p. 333. Academic Press, New York, 1958.
2. C. W. Gehrke, W. M. Lamkin, D. L. Stalling, and F. Shahrokhi, Biochem. biophys. Res. comm. 19:328, 1965.
3. A Meister, Biochemistry of the Amino Acids, 2nd ed. Academic Press, New York, 1965.
4. E. Bayer, K. H. Reuther, and F. Born, Angew. Chem. 69:640, 1957.
5. A. Liberti in: Gas Chromatography 1958, ed. D. H. Desty, p. 341. Academic Press, New York, 1958.
6. A. Zlatkis and J. F. Oro, Anal. Chem. 30:1156, 1958; A. Zlatkis, J. F. Oro, and A. P. Kimball, Anal. Chem. 32:162, 1960.
7. M. Severin and M. Renard, Rivista Ital. Sost. Grasse, No. 12: 649, 1963.
8. B. Halpern, J. W. Westley, I. von Wredenhagen, and J. Lederberg, Biochem. biophys. Res. comm. 20:710, 1965.
9. G. E. Pollock and V. I. Oyama, J. Gas Chrom. 4:126, 1966.
10. E. Gil-Av, R. Charles-Sigler, G. Fischer, and D. Nurok, J. Gas Chrom. 4:51, 1966.
11. F. Weygand, B. Kolb, A. Prox, M. A. Tilak, and I. Tomida, Hoppe-Seyl. Z. Physiol. Chem. 322:38, 1960.
12. K. Blau and A. Darbre, Biochim. biophys. acta 126:591, 1966.
13. F. Weygand and R. Geiger, Chem. Ber. 89:647, 1956.
14. P. A. Cruickshank and J. C. Sheehan, Anal. Chem. 36:1191, 1964.
15. D. L. Stalling and C. W. Gehrke, Biochem. biophys. Res. comm. 22:329, 1966.
16. A. Darbre and K. Blau, J. Chromatog. 17:31, 1965.
17. R. Consden, personal communication.
18. K. Blau and A. Darbre, Biochem. J. 88:8P, 1963.
19. R. E. Bowman, J. Chem. Soc. (1950), p. 1349.
20. F. Weygand and H. Rinno, Chem. Ber. 92:517, 1959.
21. A. Darbre and K. Blau, Biochim. biophys. acta 100:298, 1965.
22. E. E. Schallenberg and M. Calvin, J. Am. Chem. Soc. 77:2779, 1955.
23. D. E. Johnson, S. J. Scott, and A. Meister, Anal. Chem. 33:669, 1961.

Applications of Specific Detectors
in Microanalysis by Gas Chromatography

Arthur Karmen

Department of Radiological Science
The Johns Hopkins Medical Institutions
Baltimore, Maryland

Although the high sensitivity of the detectors is largely responsible for the wide applicability of gas–liquid chromatography (GLC) in biological studies, it is not the only characteristic to consider when choosing a detector for a given analytical problem. The specificity of the detector often plays a more important part in determining whether the desired information will be obtained in a given analysis and how reliable that information will be. In this paper, several examples of applications of highly specific detection systems that demonstrate the contribution these devices can make will be described.

The detectors available for GLC vary widely in specificity. Several devices based on monitoring a physical property of the gas such as thermal conductivity, density, and sound velocity hardly distinguish among chemically different compounds. Other devices including rapid scanning mass, ultraviolet, or infrared spectrometers offer such high specificity that they are frequently usable without separating the components of a mixture. Intermediate in specificity are those detectors that respond to a particular chemical functional group.

This work was supported by U.S.P.H.S. Grant GM-11535

These devices include automatic titrators, spectrometers set
to monitor one specific wavelength or ratio of wavelengths, and
detectors based on chemical or physical reactions specific for
single elements or simple groups of elements such as halogens,
sulfur, or phosphorus. In GLC with detectors of intermediate
specificity, compounds are still identified by their retention
times on the GLC column, with a degree of certainty that
depends just as much on the resolution of the column as when
nonspecific detectors are used. However, because of the specificity of the detector, the compounds with particular functional groups can be distinguished from others with similar
retention volumes on the column. Thus, the specificity of the
detector can complement the resolving power of the column by
offering information about the identity of various compounds.
Several applications of one of these devices, a specific detector
for halogens and phosphorus based on flame ionization,
demonstrate aspects of the utility of these devices that have
so far not been exploited.

METHODS AND MATERIALS: SPECIFIC FLAME
IONIZATION DETECTION OF HALOGENS
AND PHOSPHORUS

It has been observed that the sensitivity of a hydrogen
flame ionization detector to compounds containing halogen or
phosphorus is enhanced if a source of alkali metal salt is
heated in the flame [1]. The reaction between a halogen compound, and a heated alkali metal salt also forms the basis of
two other halogen detection methods [2, 3]. The mechanism of
this reaction was studied to determine whether the increased
current in the presence of halogen could be attributed to an increase in the ionization of alkali metals already present in
the flame or whether halogen and phosphorus increased the
vapor pressure of alkali metal and thus made more of it
available for ionization. Although an increase in the ionization
of the alklai metal already present by halogens was not ruled
out, it was found that halogens significantly increased the vapor

pressure of the alkali metal. This finding made possible the development of a specific detector for halogens and phosphorus [5, 6].

The hydrogen flame ionization detector is known to be insensitive to products of the combustion of organic compounds such as carbon dioxide, carbon monoxide, and water. Since the compounds that are ionized in a hydrogen flame include only organic compounds, alkali and alkaline earth metals, only these metals remain detectable by flame ionization once the gases have been subjected to combustion. A wire mesh screen that had been treated with a solution of an alkali metal salt was mounted above the flame of a hydrogen flame ionization detector. The presence of halogens or phosphorus in the flame gases increased the rate of volatilization of the alkali metal from the screen.   The vapor pressure of the alkali metal was monitored by flame ionization in a second hydrogen flame mounted just above the heated area of the screen. The electrical conductivity of the lower flame was a function of the concentration of organic material burning in it just as is the electrical conductivity of the flame of the conventional hydrogen flame ionization detector.   Since no unburned material reached the upper flame, its electrical conductivity was a function of the concentration of alkali metal vapor in the gas above the screen.

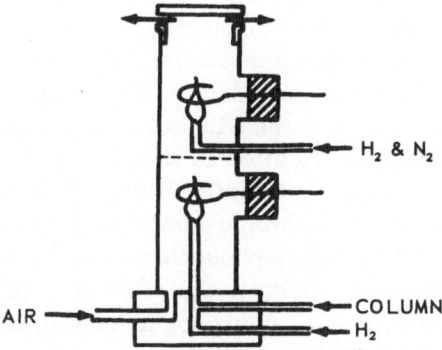

Fig. 1. Schematic of dual hydrogen flame ionization detector specific for halogens and phosphorus.

Changes in it thus reflected only the concentration of halogen or phosphorus in the gas burning in the lower flame.

The detector cell has been described in detail [5] (Fig. 1). A screen constructed of 52 mesh platinum wire gauze (Fisher Scientific Company) was mounted in a conventional hydrogen flame ionization detector approximately 15 mm above the flame jet. A second hydrogen flame detector was installed directly above this screen. Both flames were provided with electrodes for measurement of their respective electrical conductivities. Different alkali metal salts have been used to treat the screen including sodium chloride, sodium sulfate, lithium chloride, potassium chloride, rubidium chloride, and cesium chloride. Treatment consists simply of dipping the screen in an aqueous solution of the alkali metal salt and drying it in a flame.

The electrical conductivity of both flames was recorded simultaneously in each analysis so that the "total" organic material was recorded as well as the halogen or phosphorus.

APPLICATIONS

Trace Analysis for Phosphorus Compounds

In "trace analyses," specific compounds of interest present in relatively small quantities must be distinguished from much larger quantities of other compounds. It is usually necessary to subject the sample to extensive purification procedures by solvent extractions and other forms of chromatography prior to injecting the sample into the GLC column. When the detector is specific, less pre-purification is necessary. Figure 2 shows an analysis of a solution of methyl esters (10 μg) containing trace quantities of a chlorine-containing compound, butyl chloroacetate, and two organophosphorus compounds, triethyl- and tributyl phosphates. Despite the similarity in the retention times of the chlorine- and phosphorus-containing compounds and those of the methyl esters, the compounds were easily distinguished on the upper flame record because of the high sensitivity of the upper flame detector to halogens and phosphorus

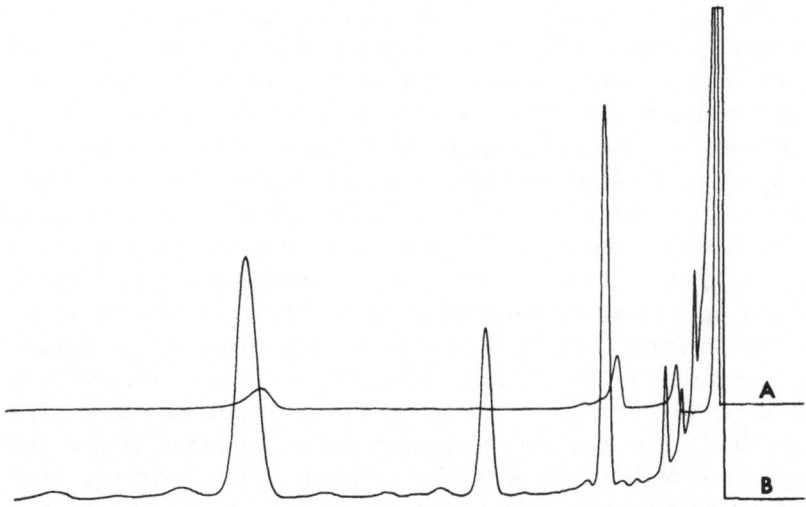

Fig. 2. Analysis of a solution containing 10 μg of a mixture of methyl esters, 150 mg of butyl chloroacetate, 18 ng each of triethyl and tributyl phosphates. There was no response in the upper flame (upper record) to the methyl esters.

and its insensitivity to the methyl esters. Analysis of a relatively large sample (by GLC standards) of methyl esters (Fig. 3) showed that the sensitivity of the upper flame detector to phosphorus was five orders of magnitude greater than that to methyl ester.

In these trace analyses, the lower flame record provides several kinds of useful information. By its continuous record

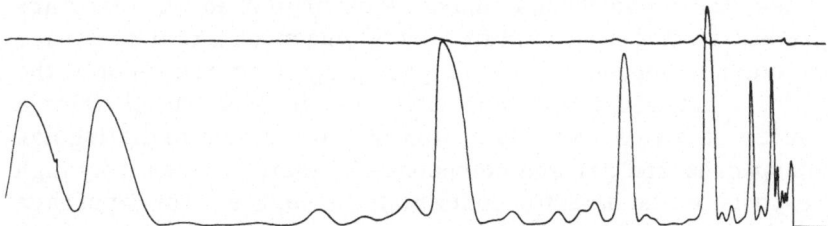

Fig. 3. Analysis of 1 μliter of a mixture of methyl esters of fatty acids. The response of the lower flame was recorded at 3×10^{-7} A full scale; the response of the upper flame at 3×10^{-8} A full scale. Approximately 0.25 mg of methyl palmitate, the large peak in the center of the record, was barely detectable on the upper record.

of the concentration of total organic material in the flame gases, it helps in determining when, and to what degree, the upper flame response is attributable to compounds that do not contain halogen but that are present in relatively high concentrations. If known organic compounds are present in the mixture, or if these are added intentionally, then the lower flame record provides a useful check on the retention times of these compounds during each analysis. This also serves to increase the certainty of identification of trace compounds on the record of the upper flame. Finally, by providing a complete analysis of the sample, it helps in monitoring the efficiency of "clean-up" or pre-GLC purification procedures. Thus, although it is possible to record only the output of the upper flame, with its specificity for halogens and phosphorus, and thus to save the expense of one channel of amplification and data recording, much important data will be lost if this is done.

The most important function of the lower flame in this detector is to combust the sample and thus to render organic material that does not contain phosphorus or halogen undetectable by the upper flame. Much of the specificity of the device is thus attributable to this function. It is possible to heat a source of alkali metal in a single flame detector so that the metal passes through the flame and is ionized. Although the sensitivity of this flame detector to phosphorus and halogen containing compounds is then enhanced, its sensitivity to other compounds is retained [6]. Even though the sensitivity to phosphorus compounds is increased three hundred or more times, these compounds cannot be distinguished when they are present as traces in mixtures containing much larger quantities of other compounds. It is possible, of course, to split the effluent between two flame detectors, in only one of which a source of alkali metal is heated, and to attempt to distinguish phosphorus and halogen compounds by their differentially high response in the detector containing the source. This approach is also ineffective when the phosphorus or halogen is present in very small quantities because of the difficulty of distinguishing a small increment on top of a large signal current. The upper—lower flame arrangement is much more effective in trace analyses because of its greater specificity.

There are many other examples that demonstrate the importance of the specificity of the detector in trace analyses.

Goulden, Goodwin, and Davies [7] compared the results of analyses of chlorine-containing pesticides performed with an electron capture detector and with a General Electric Co. thermionic halogen element [2]. They reported appreciably better results with the halogen detector despite its lower sensitivity because it was less subject to interferences. These improved results were obtained despite the fact that this detector responded to some degree to nonhalogen-containing compounds.

Specific Detectors in Derivative Analyses

Halogen-containing derivatives of a number of chemical classes of compounds can be prepared and analyzed by GLC. When these derivatives are detected by the dual-flame halogen detector, the upper detector output responds in proportion to the number of functional groups that react with the derivatizing reagent. The detector can thus distinguish compounds that have these groups from others that have similar retention volumes, a property that is useful in trace analyses. This responsiveness to specific functional groups serves as an adjunct to group separations by ion exchange and other forms of chromatography prior to GLC. Comparison of the upper and lower detector outputs also yields information on the ratio of the number of functional groups to the size of the remainder of the molecule. This information can supplement the measurement of retention times as an aid in identifying compounds.

This general approach has now been applied to the analysis of mixtures of alcohols and amines as their corresponding dichloroacetates and dichloroacetamides, to short and long chain monocarboxylic acids and short chain dicarboxylic acids as 2-chloroethanol esters, and to aldehydes as chloroethylacetals.

N-dichloroacetamides were prepared by reacting the corresponding amines with dichloroacetic anhydride in the presence of pyridine. The amines 3 to 8 carbons long yielded derivatives that could be chromatographed under conditions similar to

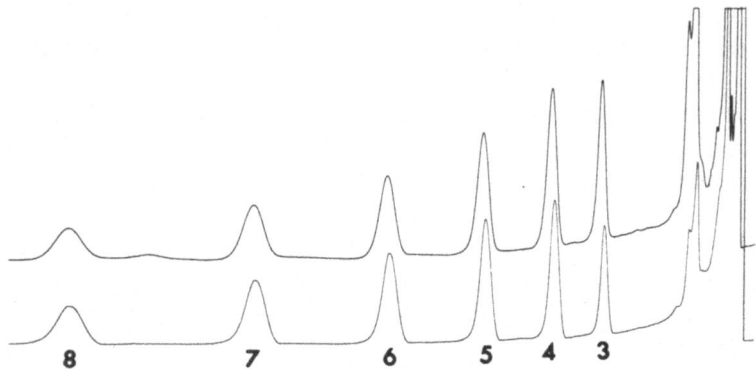

Fig. 4. Analysis of a solution containing about 20 μg of a mixture of N–dichloro-acetamides of short chain amines from thee to eight carbons long. The column was ethylene glycol adipate 14% w/w on Chromosorb W at 180°C. The upper record, that of the upper flame detector, is specific for halogen; the lower record, that of the lower flame, is that of a conventional hydrogen flame detector. Both were recorded at 3×10^{-8} A full scale.

those generally employed for the analysis of long chain fatty acid methyl esters: a column packed with ethylene glycol adipate on Chromosorb W, 10% by weight, at 180°C (Fig. 4). The relative responses of the lower flame detector to these amines were as predicted from the well-known responses of the hydrogen flame detector [8]. The responses of the upper flame correlated well with the number of molecules present. When a homologous series of amines was analyzed, the ratio of the responses of the upper to lower flames decreased with increasing chain length, as expected from the decreasing ratio of amino groups to methylene groups on the molecules. GLC of the dichloroacetates of short chain alcohols at a column temperature of 120°C yielded a record almost identical with that of the dichloroacetamides shown in Fig. 4. Trichloro-acetates of short chain alcohols, prepared by reacting these compounds with trichloroacetic anhydride, also behaved simi-larly (Fig. 5).

Chloroethyl derivatives of monocarboxylic and dicarboxylic acids and aldehydes were prepared by reacting these compounds with 2-chloroethanol in the presence of 2% sulfuric acid as catalyst [9]. Esterification and acetal formation were complete after three hours at 60°C. GLC was performed on the EGA column at from 110°–190°C depending on the chain length of the derivatives. The response of the lower flame per unit weight of short chain monocarboxylic acid was as predicted for those acids four or more carbons long. Acetic, propionic, and butyric esters yielded consistently lower results for which there was no apparent explanation. The response of the upper flame was proportional to the number of moles for all acids above butyric in chain length. As in the case of the amides, the ratio of responses of the upper to lower flames decreased with increase in chain length of the ester (Fig. 6). The larger molecules, with more methylene groups per carboxyl and, therefore, per chlorine, evoked a greater response per unit weight in the lower flame as well as a lower response in the upper flame.

The predicted response to dicarboxylic acids was also observed. The ratio of responses of the two flames (upper: lower) to dichloroethyl succinate was 0.8, compared with 0.25 for chloroethyl decanoate. The predicted ratio (upper: lower) for the succinate diester is proportional to: the number of

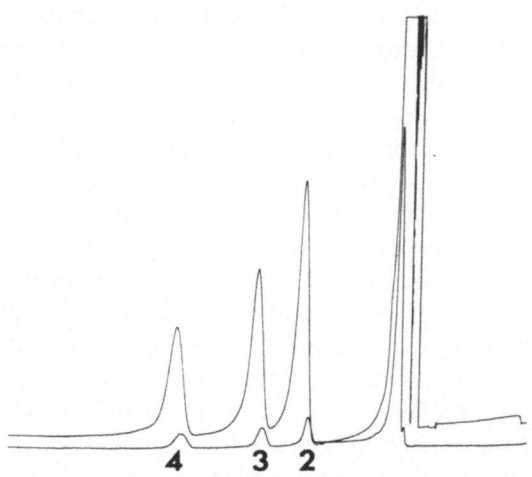

Fig. 5. Analysis of about 10 μg of a mixture of trichloroacetates of ethanol, propanol, and butanol. The upper record is that of the upper flame detector which is specific for halogen; the lower record is that of the lower flame. Both were recorded at 3×10^{-8} A full scale (122°C EGA).

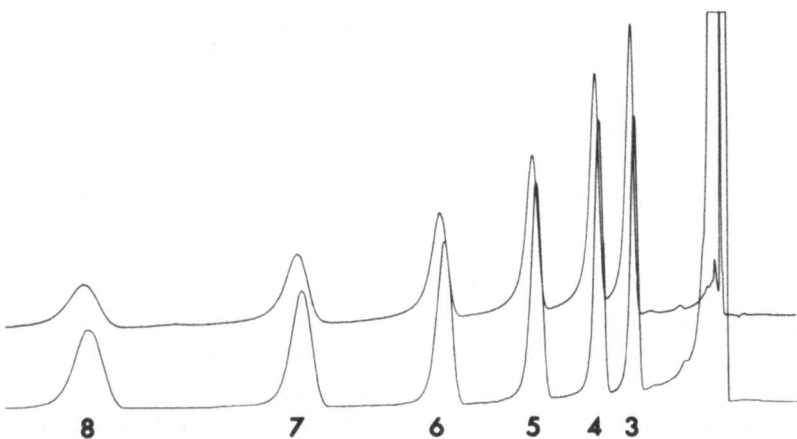

Fig. 6. GLC of chloroethyl esters of short chain fatty acids prepared from a mixture containing approximately equal weights of the various acids. The upper record was the output of the upper, halogen-sensitive flame detector; the lower record was the output of the lower, conventional flame detector. The full-scale sensitivity was 3×10^{-8} A on both records (150°C EGA). The sample injected was derived from approximately 10 μg of each acid.

moles × 2Cl/number of moles × 6 carbons. The predicted ratio for the decanoate ester, derived similarly, was 1/11. The ratio for succinate divided by the ratio for decanoate is thus 11/3, or 3.67. The quotient of the ratios actually found was 3.2.

The retention times of chloroethyl derivatives of dicarboxylic acids and aldehydes were similar to those of monocarboxylic esters of appreciably greater chain length. When they were all present in the same mixtures, the derivatives of the aldehydes and dicarboxylic acids were easily distinguished on the dual flame record not only because of their higher halogen content but also because of the relatively lower response of the lower flame. In an analysis of a mixture of chloroethyl derivatives prepared from a commercial facial soap, three kinds of compounds could be distinguished (Fig. 7). The sample was prepared by first converting the fatty acid salts to methyl esters and then incubating the methyl esters with chloroethanol-H_2SO_4 as described. In one class of compounds were the unreacted methyl esters (labeled "m" in the figure), identified by their retention times as well as by their failure to elicit responses in the upper flame. In the second class of compounds were the chloroethyl esters of the fatty acids, detected by both upper and lower flames. These esters were in the same relative proportions as the methyl esters in the original mixture. The third class of compounds (labeled "E" in the figure) had higher ratios of upper-to-lower flame responses than the chloroethyl esters and were possibly acetals or diesters of dicarboxylic acids. Several of these compounds would ordinarily have been identified as unsaturated or branched chain esters on the basis of their retention times.

SUMMARY AND DISCUSSION

In trace analyses by GLC, a specific detector is often more useful than a more sensitive, nonspecific detector. It permits small quantities of compounds of interest to be dis-

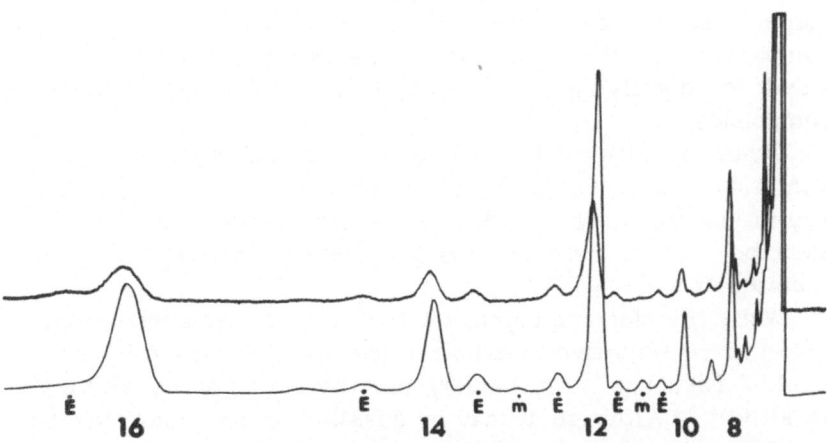

Fig. 7. GLC of a mixture of chloroethyl derivatives prepared from a sample of a commercial facial soap on the 14% EGA column. The upper record was the output of the upper halogen-sensitive flame detector; the lower record was the output of the lower, conventional flame detector. Full-scale sensitivity on the upper record was 10^{-9} A; on the lower, 3×10^{-9} A full scale. The peaks labeled 8, 10, 12 were chloroethyl esters; those labeled "m" were methyl esters; those labeled "E" contained more halogen per unit weight than chloroethyl esters of monocarboxylic acids.

tinguished from much larger quantities of other compounds with similar retention times. This is exemplified in analyses of trace quantities of phosphorus compounds in mixtures containing much larger quantities of methyl esters performed using a highly specific phosphorus detector. The advantages offered by specific detectors can also be extended to other classes of compounds by preparing derivatives containing the element or functional group to which the specific detector is sensitive. Examples presented here include analyses of chloroethanol derivatives of organic acids and aldehydes, and of chloroacetyl derivatives of amines and alcohols using a specific halogen detector. The dual-flame detector used in these analyses provided two outputs: The upper flame response was proportional to the number of moles (or rather, equivalents) of carboxyl, carbonyl, hydroxyl, or amino group, and the lower flame response was proportional to the total quantity of organic material as seen by the usual hydrogen flame detector. Both outputs provided useful information. In analyses in which the components of interest were well separated from each other on the column, as in the analyses of the fatty acid esters of soap, comparison of the response of the upper and lower flames aided in identifying and distinguishing different classes of compounds.

Other specific halogen detectors could, presumably, be similarly used. Radiation detectors have also been used this way in "isotope derivative" methods of trace analyses [10]. The electron capture detector has been used in analyses of haloacetates of sterols [11].

While the electron capture detector is appreciably (perhaps 100×) more sensitive to haloacetates than the halogen detector, it also responds to a variety of other compounds with high sensitivity. Although it may be possible to use other specific detectors in conjunction with a hydrogen flame to achieve some of the results obtained with the dual-flame halogen detector, the electron capture detector is not quite as suitable as some others because of its nonspecificity. High specificity, as exemplified by the dual-flame halogen detector, rather than high sensitivity is more useful for this purpose.

REFERENCES

1. A. Karmen and L. Giuffrida, Nature 201:1205, 1964.
2. C.W. Rice, U.S. Pat. 2,550,498 (April 24, 1951).
3. C.C. Anthes, U.S. Pat. 2,779,666 (January 29, 1957).
4. P.J. Padley, F.M. Page, and T.M. Sugden, Trans. Faraday Soc. 57:1552, 1961.
5. A. Karmen, Anal. Chem. 36:1416, 1964.
6. A. Karmen, Gas Chromatog. 3:336, 1965.
7. R. Goulden, E.S. Goodwin, and L. Davies, Analyst 88:951, 1963.
8. J.C. Sternberg, W.S. Galloway, and D.T.L. Jones, in: N. Brenner, J.E. Callen, and M.D. Weiss, eds., Gas Chromatography, Academic Press, New York, 1962, p. 231.
9. A. Karmen, J. Lipid Research, Vol. 8, 1967, in press.
10. A. Karmen, I. McCaffrey, and B. Kliman, Anal. Biochem. 6:31, 1963.
11. R. Landowne and S.R. Lipsky, Anal. Chem. 35:532, 1963.

INDEX